THE NATURE OF THINGS ON SANIBEL

REVISED EDITION

A Discussion of
The Animal & Plant Life
of Sanibel Island
With a Sidelong Glance at
Some of Their Relatives
Elsewhere

George R. Campbell

Drawings by Molly Eckler Brown

Pineapple Press, Inc.
Sarasota, Florida

Pineapple Press, Inc.
P.O. Drawer 16008
Sarasota, Florida 34239

Library of Congress Cataloging-in-Publication Data

Campbell, George R. (George Robert), 1918-
The nature of things on Sanibel.

Includes index.
1. Natural history—Florida—Sanibel Island.
2. Ecology—Florida—Sanibel Island. I. Title.
QH105.F6C32 1988 508.759'48 87-36112
ISBN 0-910923-47-7

First revised edition, 1988
10 9 8 7 6 5 4 3 2
First edition published in 1978

Printed and bound by Arcata Graphics, Kingsport, Tennessee
Designed by Molly Eckler and Jean W. Culpepper

Let us beware of needlessly destroying even one of the lives — so sublimely crowning the ages upon ages of evolving; and let us put forth all our efforts to save a threatened species from extinction; to give hearty aid to the last few individuals pitifully struggling to avoid absolute annihilation.

The beauty and genius of a work of art may be reconceived, though its first material expression be destroyed; a vanished harmony may yet again inspire the composer; but when the last individual of a race of living beings breathes no more, another heaven and another earth must pass before such a one can be again.

William Beebe, 1906

*This book is dedicated to my wife,
Jean, and to Duane, whose encouragement
and support made it possible.*

ACKNOWLEDGMENTS

It would have been foolhardy for me to attempt this book without the cooperation of Charles LeBuff and his knowledge of local flora and fauna. I cannot overemphasize the help that he has willingly extended to me and the endless hours spent reading my drafts.

Other local naturalists who have been very helpful are Richard Beebe, Emily Shane, Robert Watterson, George Weymouth and Richard Workman.

I express my deep appreciation to the following scientists for their help in verifying data and references, aiding in the accumulation of information and contributing in ways too numerous to mention: Dr. Morton S. Biskind, David G. Campbell, Dr. Howard Campbell, Dr. Archie Carr, Dr. Elias Cohen, Dr. Frank C. Craighead, Sr., Dr. Earl Frye, Jr., Dr. Jean W. Gentry, Delbert Hicks, Dr. Gordon Hubbell, Dr. James N. Layne, Thomas M. Missimer, Dorothy Raeihle, Dr. Harrison B. Tordoff, Dr. F. G. Walton Smith and Dr. William Webb.

I want to express my appreciation to Jean Culpepper, my editor, and to Gloria Berry, Dr. Marcia K. Miller, and Laura and William Riley for their helpful editorial comments on the manuscript.

Appreciation is also extended to Molly Eckler Brown for her excellent drawings and to the following people who granted us permission to reproduce their Molly Eckler Brown art: A. Cherbonnier, Don and Grace Whitehead, Dick and Judy Workman, David A. Pugh, Edward and Jane Cissel, Ikki Matsumoto, Jean Campbell, Peter Valtin, Gregory O'Berry, Island Printing Center of Sanibel, John and Mary Eckler.

My thanks to all those photographers whose work appears in this book, and to Ann L. Winterbotham for her fine mapwork.

My deepest gratitude goes to Duane White without whose substantial support and encouragement these pages would never have reached the printed stage.

Thanks also to Don Whitehead and *Island Reporter,* in which publication much of this material appeared originally in different forms.

And finally, thanks to my boss, Cleveland Amory, for his encouragement and support.

George R. Campbell

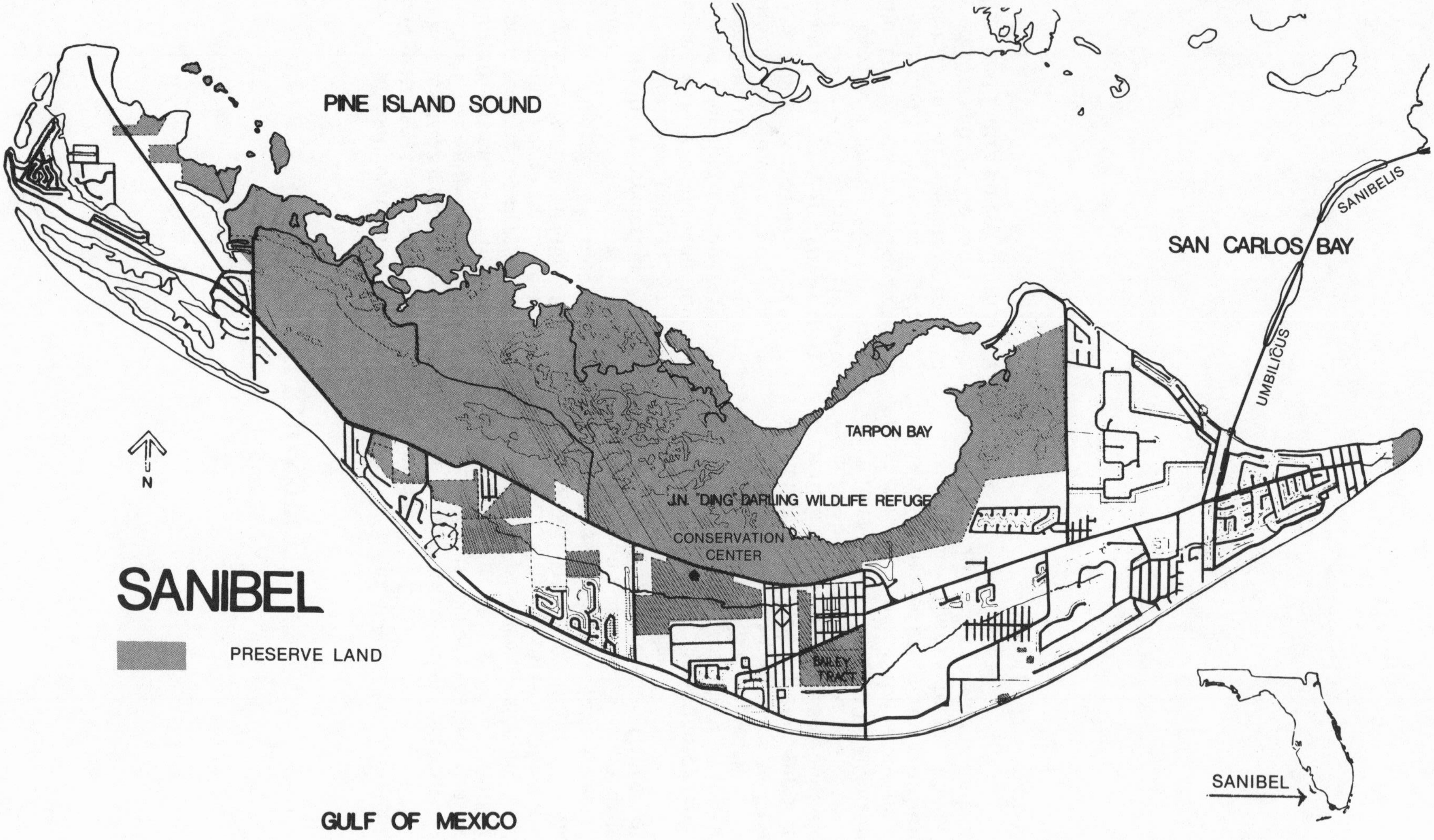

SANIBEL

PRESERVE LAND

PINE ISLAND SOUND

SAN CARLOS BAY

SANIBELIS

UMBILICUS

TARPON BAY

J.N. "DING" DARLING WILDLIFE REFUGE

CONSERVATION CENTER

BAILEY TRACT

N

GULF OF MEXICO

SANIBEL

CONTENTS

ILLUSTRATIONS

INTRODUCTION

by CLEVELAND AMORY
President, The Fund for Animals

I once asked George Campbell, who works for The Fund for Animals, what was his definition of the word "boss". "Someone," he replied, "who can take directions quickly and accurately."

So, when Mr. Campbell ordered me to do an introduction for this book, I, of course, hopped to it. However, as boss, I feel I have a few small rights. And one of these rights I hereby exercise. In the first sentence of his foreword, George declares that he hoped to avoid in his book the use of the word "unique". He also counselled me in my introduction to avoid the word.

I of course agreed. George Campbell is a unique fellow who lives in a unique place, and he has written a unique book.

I have been there. And to go to Sanibel in the company of George and Jean Campbell is to understand the meaning of what he has, in his foreword, also written:

> *Here we have in microcosm the environmental problems of the whole world. Here is a valuable and graphic example small enough to hold in your hand, figuratively, and examine under the microscope. The principles and problems here observed can be extrapolated to most other areas in the world, where similar problems are abundant, terrifyingly serious, often terminal . . .*

Through these pages you will explore with George a world of alligators and armadillos, bobcats and dolphins, panthers and pelicans, ospreys and opossums, iguanas and egrets, raccoons and river otters, roseate spoonbills and marsh rabbits, horseshoe crabs and a host of other creatures. It is typical of George that not only does he never harm these but even when he writes of fishing, the fishing feat of which he is most proud is that of having caught the world's record tarpon — the *smallest* ever taken — just one inch long, which he of course saved, not destroyed.

You will read here fascinating stories of familiar animals — like whales and dolphins. But the remarkable thing about this book is that you will also read stories about unfamiliar animals — even animals nobody but George and Jean seem to love. Here you will read, for example, personal accounts of "Bismarck" and "Twiggy" and "Marshall". And who are Bismarck and Twiggy and Marshall? They are alligators. There is no man living, in my opinion, who knows more about alligators than George. You will also read about "Beelzebub." And who is Beelzebub? A gopher.

You will meet people too in this book — some people almost as fascinating as the animals, like the island's lady Dr. Dolittle. George of course is much too modest to call himself a Dolittle. But the fact is when I visited George and Jean's greenhouse, I found more than half a hundred animals in various stages of being Dolittled toward recovery and eventual return to the wild.

"What are you doing?" I asked George one day, when I telephoned him from a great distance. "I'm sitting on my patio," he said, "trying to think of a reasonable excuse for including the story of the Kirtland's warbler in a book on Sanibel's flora and fauna."

There and then I knew that whatever excuse George came up with, he would also come up with my kind of book.

FOREWORD TO SECOND EDITION

Since this book first appeared in 1978, a lot has happened to Sanibel. Most of what has happened has had a negative effect on the wild animals and plants that visitors and residents alike found so fascinating and valuable.

Few beneficial things have occurred. Outstanding among those few were 13 animal crossings that were installed beneath the Sanibel-Captiva Highway. These structures represent a significant pioneering effort on the part of Ann L. Winterbotham, then chair of the Sanibel Planning Commission. Not uncommon in Britain, these were, we think, among the very first to be proposed in this country.

In this foreword to the second edition, I shall, chapter by chapter, enumerate and discuss the pros and cons of some of those happenings that have affected species discussed in this book.

Chapter 1 - *THE RIVER OTTER*

The mother Otter who gave her life on the San-Cap Highway in April 1975 was but one of a dozen or so since that time. Careless motorists continue to destroy Otters and the Otter population shows a substantial decline as a consequence.

Another factor contributing to the decline in the Otter population is the silting in of mosquito ditches. Some years ago, the Army Corps of Engineers decreed that filling in the wetlands was not to be allowed. Well, if you dig in the wetlands, you have to put the fill somewhere, and this rule effectively halted the creation of new mosquito ditches and the maintenance of the old ones, i.e., keeping them open and free. While this policy has a lot of good aspects that aid in the preservation of Sanibel wildlife, it has a serious negative effect on Otters. Our Otters have enjoyed very large linear ranges. They coursed many of the mosquito ditches in the past, and still do where they can get through. But in many places they are inhibited by the silting in and blocking of the ditches.

In the spring of 1986, an Otter was killed after being hit by traffic on McGregor Boulevard near the Sanibel toll booth. This tragic occurrence was witnessed by many people who were coming to work on Sanibel that morning. It caused a furor at the time, but it soon died down, and many drivers are back at their careless worst — slamming away at our wildlife.

It is sad to contemplate, but I fear that the decline in our Otter population presages the complete annihilation of the species on Sanibel within the next few years. Of course, the huge tract of the J.N. "Ding" Darling National Wildlife Refuge might help preserve the island population, but in the last couple of years I have seen fewer Otters there and elsewhere on the island. I believe all Otters on the island are exposed to traffic hazards.

Chapter 2 - *THE ISLAND'S ONLY MARSUPIAL*

I am happy to report that, as expected, the Opossum is doing quite well on Sanibel. A hundred or more are traffic fatalities each year but these prolific creatures produce large litters of young. Even when a mother with young is killed, her young are often taken to CROW (Care and Rehabilitation of Wildlife, Inc.), where they are raised and released into the wild.

Chapter 3 - *THE FLORIDA PANTHER*

Since writing the original chapter on the Florida Panther, many things have taken place. First, the school children of Florida, unwisely I think, selected the Florida Panther to be the Florida State Animal, because it is much loved and admired and so close to extinction. Perhaps it would have been better had they chosen the Otter or some other treasured member of the wildlife community. Far too often serious attention is paid to saving an animal species or subspecies, as in this case, only after it is seriously endangered and perhaps no longer salvageable.

Aside from road kills, Seminole hunters, habitat destruction and disease, there is another threat to the Florida Panther: hybridization with alien subspecies of the Cougar from other areas. Quite a number of California Cougars and some South American ones have been released in Florida. There is no doubt in this writer's mind that they have hybridized with the Florida form. If there are any pure Florida Panthers left, they may very well be the two, three or four that share the huge range that is formed like a donut which completely encircles Fort Myers, including the chain of barrier islands which includes Sanibel and Captiva. Since originally publishing this book, the writer has gathered sighting data from many sources and believes that the Panther today still ranges in a circle around Fort Myers.

Chapter 4 - *DEMISE OF THE WHALES*

False Killer Whales are still beaching themselves in the region. In May of 1986, a pod beached at Fort Myers Beach and some other beaches to the south. Most, perhaps all, died.

A new threat to our porpoises was proposed in mid-1986 when one of the island resorts proposed the capture of eight young *Tursiops* for public display. Of course, all local environmentalists opposed this plan and the resort wisely cancelled it. But this instance only serves to show that vigilance is absolutely necessary in all conservation efforts. Like that famous ballgame, the battle is "never over until it is over." By the same token, changes favorable to cetaceans have been made in the decisions of the International Whaling Commission, where recently almost a total ban on all commercial whaling was voted for. But before the year was out, Norway, Japan and the Soviet Union were violating the letter and spirit of the decision of the International Whaling Commission.

Chapter 5 - *THE ANCIENT ALLIGATOR*

The Alligator on Sanibel today is in a very ambiguous position. Since this book first appeared, the state of Florida has assumed total responsibility for the management of the species, having taken over from the United States Department of Interior. I find it deplorable that the United States government is unable to carry on with the range-wide multi-state management of the species. Instead the Feds too often toss the management responsibility of a species back to the states. In the present case, in the remaining nine-state range of the American Alligator there are about that many different kinds of management approaches. In Florida it is pretty bad, maybe the worst. Florida has brought former Alligator poachers out from the woodwork, licensed them, and given them a new lease on life by authorizing them to kill "nuisance" Alligators. The killer gets 70% of the value of the hide after it is auctioned off by the state, and he gets all of the meat. Sanibel still remains the only exception in the state of Florida where nuisance Alligators are not killed but

are removed by licensed handlers. Until mid-1986, the handlers were members of the Southwest Florida Regional Alligator Association which I founded a number of years ago. From mid-1986 to the present the job has been handled by state licensed Alligator handlers under City of Sanibel control.

Actually, no appreciable increase in licensed handlers really needed to have taken place, because the practice of not feeding Alligators according to law has reduced the number of true nuisance Alligators greatly. Consequently, the second echelon of Alligator complaint respondents that I selected and put in force in April 1986 precludes the necessity of capturing animals in most cases. This second echelon of knowledgeable people responds to Alligator complaint calls, explains the status of the species, and tells people how to co-exist successfully with this valued animal on Sanibel without threat to themselves, their dogs, or to the Alligators themselves.

Chapter 6 - *THE ARMADILLO*

The Armadillo is expanding its population and getting along quite well, perhaps too well. The litters of identical quadruplets seem to insure the species' continued successful survival on Sanibel. There would be a lot more of them if they could conquer one strange and lethal habit. Whenever a car passes directly over one in an effort to straddle the animal in the road, the animal does his best to commit suicide by jumping straight up into the underside of the car.

The embryos are marvelously tenacious of life. I found a freshly road-killed pregnant mother on the highway with four two-inch embryos forced from her body and lying on the pavement. I took them home with me not at first realizing they were still quite alive. When I washed them off in a bucket of water preparatory to preserving them, they began to swim quite vigorously!

Many deplore the success of the armadillo. It eats the eggs and young of lizards, snakes, and ground-nesting birds. It digs up garden plants too. I'll not condemn this strange and fascinating animal. I prefer to enjoy it.

Chapter 7 - *MARSH RABBIT*

On Sanibel, the Marsh Rabbit does not seem to undergo the same population fluctuations that occur elsewhere in its range. Perhaps this is due to there being few remaining natural predators and the animal's ability to evade the common "surrogate predator" so successfully — i.e., it is seldom smashed on the road by passing cars. So almost any time of the year these attractive creatures can be observed.

Chapter 8 - *J.N. "DING" DARLING NATIONAL WILDLIFE REFUGE*

The "Ding" Darling Refuge has been managed with greater skill than in some periods in the past. Ron Hight has been in the top position for several years in the mid-eighties, and under his administration the Refuge has been able to halt the baiting of crabs and the feeding of Alligators, with appreciable benefits. The surface of the Memorial Drive through the Refuge has been greatly improved, with mixed results. Now those who are inclined to speed can speed faster. On the other hand, most good citizens can make their visits more comfortable and pleasant. The Gasparilla Trail has been more realistically described as the Indian Mound Trail and is greatly improved with boardwalks throughout. Strategically placed hand railings serve to protect some increasingly rare epiphytes.

The Visitors Contact Center is well developed now, and has adequate dioramas depicting local wildlife. There is also a small room where films are shown regularly.

Chapter 9 - *SANIBEL-CAPTIVA CONSERVATION FOUNDATION*

The SCCF Sanctuary is no longer the wild place it was in 1978. Broad paths, cleared areas and occasional benches make it less a sanctuary. In a way the SCCF mirrors what has happened to Sanibel. In the end it may simply be the park in the midst of a bustling resort.

Chapters 10 and 11 - *INSECT CONTROL*

I am happy to report that efforts to calm down the activities of the Lee County Air Force have indeed borne fruit. The Lee County Mosquito Control District still has ancient DC-3s that fly low over Sanibel occasionally, spewing quantities of dangerous organophosphate chemicals, mostly baytex and malathion. But the incidence of this application is much less frequent than it was when these chapters were first written. The Lee County Mosquito

Control people are employing more environmentally sound techniques, and in their research department they are studying many more new approaches. There have been no serious outbreaks of St. Louis encephalitis although sentinel chickens have turned up positive on occasion. The hyped-up P.R. of the District still emphasizes disease-carrying mosquitoes, while we all know that the mosquito control effort exists to support the tourist industry and the ringing of cash registers, and that disease has little to do with it at all. Only one mosquito is involved with disease, *Culex nigripalpus*, and we know, as stated above, there has been no human SLE outbreak in spite of indications that the virus is present in this region.

Chapter 12 - *CATTLE EGRET*

This wonderful, vigorous species is forever expanding its range. The bold creature even comes into populated areas and involves itself in harvesting insects during home lawn-mowing activities.

Chapter 13 - *SHARKS AND RAYS*

I still haven't been swimming in Shark-populated waters since writing this chapter some years ago. I still keep records of Shark attacks, and I'm still concerned about their sharp teeth. In October 1986 a small child was severely bitten by a Shark while bathing in shallow water with her family on Sanibel Beach. She suffered very serious injuries which will affect her whole life. Having studied the conditions surrounding a couple of Alligator bite incidents, I no longer make the comparison that it is safer to swim in freshwater with Alligators than in saltwater with Sharks. I think both are risky. The only time I ever go in the sea is to capture that stray Alligator who, in confusion, sometimes crosses the beach dune and enters the Gulf. In such a condition it is quite safe to approach the confused animal, noose him if I can, and haul him back into the freshwater slough. On those rare occasions, I tend to forget that I am indeed afraid of Sharks.

The Sting Rays are still abundant at certain times of the year, as are human casualties from their sharp barbs.

Chapter 14 - *SNAKES ALIVE*

The snakes of Sanibel are suffering from human overpopulation, habitat destruction, overdevelopment, and too many cars speeding on too many good roads. Today it is not easy to find Ribbon Snakes, even on rainy nights. You can drive ten miles and see two, where in the past you might have seen twenty or thirty.

The Green Water Snake is believed to be extinct on Sanibel today. The Florida Brown Snake has been seen no more than three or four times over the years. Water Snakes are still present in the freshwater slough, as are the Mangrove Water Snakes in the salt and brackish water areas.

Indigos are present. More people recognize the animal and understand that it is protected by state and federal law. Education has proven valuable in this instance, because more residents recognize it and tend to take care of it.

The Southern Black Racer is still quite abundant, and each year large numbers of its patterned young hatch from eggs and get stuck in screened porches, swimming pools, gardens and all manner of places. These little animals always cause a great deal of consternation because the "experts" insist that they are Pygmy Rattlesnakes, which they are not. The Pygmy Rattlesnake has never been found on Sanibel.

The Eastern Diamondback Rattlesnake is still occasionally seen, but I feel that the pressure is great and that the species is disappearing.

The Eastern Coral Snake is still plentiful and, due to its cryptic habits, is getting along quite successfully. The Coral Snake mimic, the Scarlet King Snake, has not been seen since the first one that is described in this book.

Chapter 15 - *LIZARDS*

This writer was terribly wrong when he thought that the Six-lined Racerunner, here termed "streaker," would do well. Actually, the creation of high filled lands for house lots was halted, and although such areas do serve as good streaker habitat, lawn and garden chemicals and fertilizers condemned these lizards. Before he left the Sanibel-Captiva Conservation Foundation in 1986, Steve Phillips told me that on spoil mounds along the Sanibel River there was still an appreciable population of this species, but I believe it is a declining form on Sanibel.

The Green Anole holds its own, and there is some controversy among herpetologists as to whether it really is stressed seriously by the Key West Anole, which is very abundant and increasing.

Iguanas are still occasionally seen. The Indian House Grecko is expanding its range from the center of distribution at the trailer park and now may be found in many parts of the island.

Recent years have not turned up additional specimens of the Eastern Glass Lizard.

Chapter 16 - *TURTLES AND TORTOISES*

In 1984 we won a victory for the Gopher Tortoise population at what is known as the "Spoonbill" property. Tortoise "experts" testified that Gopher Tortoises would do well on golf course turf, eating golf course insecticides and fertilizers. Our citizens can and did recognize intellectual prostitution when they saw it. The proposed mid-island golf course and country club was defeated by the democratic process, and the Gopher Tortoises in that region have a new lease on life. In general, Gopher Tortoises are, however, seriously stressed on the island due to traffic. The coarse grasses which are their favorite natural food grow best along road rights-of-way. Consequently the animals are constantly at risk from traffic, and I believe the population is gradually being reduced. Many people pick up injured tortoises and attempt to put them back together using various materials, the one preferred by me being silastic. Other turtles are also suffering from traffic. The large female Florida Softshell Turtles that used to come out in great abundance in the springtime to lay their eggs are now mostly dead.

The Florida Box Turtle, which in 1986 was declared to be a species of special concern by the state of Florida, is pretty well gone on Sanibel due to traffic.

Charles LeBuff, founder of Caretta Research, Inc., the sea turtle study and conservation agency, reported a good year for the Loggerhead Turtles in 1986, with 137 nests discovered. Eve Haverfield, turtle guardian *par excellence*, continues her very positive sea turtle conservation activities.

If all turtles were as smart as the Osceola or Florida Snapping Turtle in their relations with automobile traffic, things would be a lot better in the turtle world. There is an old snapper on Tarpon Bay Road near the Bailey Tract that may be seen crossing the road with certain frequency. This creature observes traffic from a safe distance, and employing a fast "high walk" dashes for the river long before there is any danger of being squashed by a car.

Chapter 17 - *EDIBLE PLANTS*

The conditions are unchanged, except that I must report that Alice Kyllo makes some two dozen kinds of preserves from Sanibel's bounty. These excellent goodies were available at the SCCF Nature Center for a number of years.

Chapter 18 - *POISONOUS PLANTS*

Nothing more need be said here except to refer the reader to this writer's book, "An Illustrated Guide to Some Poisonous Plants and Animals of Florida," where we have done an exhaustive study on this subject. This valuable book is ably illustrated by Sanibel artist, Ann L. Winterbotham, and is published by the publisher of this edition of the *Nature of Things on Sanibel.*

Chapter 19 - *HORSE FOOT*

These timeless creatures continue their age-old routines on Sanibel beaches and mud flats. Dr. Elias Cohen reports great progress in the study of the valuable blood of *Limulus* and its numerous newly discovered applications. It is now widely used as a detector of pollution in the environment, and of course in parenteral drugs of the pharmaceutical industry.

Chapter 20 - *THOU SHELL NOT*

Much progress has been made in halting the over-collecting of live shells on Sanibel. In 1987 it became illegal to over-collect live shells.

Many courageous residents and devoted beachwalkers tried for years to persuade tourists that washed-up shells can be just as beautiful as those wastefully killed after being collected alive from their natural habitat. Such teachers-on-the-beach were often insulted by those who would kill our marine life. Now there is a law. Violators can be prosecuted.

Some progress has been made, some population studies are being undertaken, and it is possible that conservation efforts might keep ahead of the hordes of ever-increasing numbers of tourists who come to what now must be considered to be a resort "tourist destination" and no longer an island where the conservation effort is paramount.

Chapter 21 - *THE GRAND OLE OSPREY*

The Osprey is prospering. The nesting platforms provided years ago are for the most part still in place, and others are being added by the International Osprey Foundation, which has taken over from this writer and the Fund For Animals the management of the Osprey on Sanibel. Records show a healthy and increasing population.

Chapter 22 - *TO FEED OR NOT TO FEED*

People still feed Raccoons, occasionally are bitten, and sometimes have to take the rabies prophylactic series. Rabies in Raccoons in the Southeast is present and spreading. A few years ago, Raccoon rabies was confined to Florida and Georgia, but due to the misguided yet legally permitted export of Raccoons from Florida, including Sanibel, to northern hunting clubs, this disease is becoming widespread along the Eastern Seaboard as far as Maryland and even Pennsylvania and New York, and westward to Kentucky and nearby states. It is a mistake to have any close contact with Raccoons in Florida, and that goes for Sanibel too.

Chapter 23 - *TREE SWALLOW*

I believe the Tree Swallow is declining in numbers on Sanibel due to the fact that hundreds of southern bayberry trees have been destroyed as development has burgeoned. That is another way of saying that there is not as much winter food for the migrating Tree Swallow today as there was just a few years ago. However, in eastern Lee County there are abundant bayberries, and one can still see sky-blackening swarms of this most intriguing of our avian denizens.

Chapter 24 - *SANIBEL'S NOCTURNAL GLEE CLUB*

Concerning the toads and frogs on Sanibel, we have a sad story to tell. We still see frogs jumping in our headlights on a warm, rainy summer night. Unfortunately nine out of ten of them are poisonous Cuban Tree Frogs. On Island Inn Road sometimes the very long low leap of the Southern Leopard Frog can be detected. In that frog paradise of yesteryear, Rabbit Road, there is nothing left today. Nor are the Green Tree Frogs of Sanibel making it. The day is gone, I fear forever, for there to be thirty or forty of these beautiful jewels on the walls near your porch light, consuming the insects attracted by the light. One or two, yes, but more than that, never.

The introduced Little Greenhouse Frog manages quite well. This interesting introduction of course does not require water to breed. Its larvae are retained within the egg in damp places in the greenhouse or under a woodpile. They seem to withstand the rigors of insecticide contamination too. Perhaps they have some extra vigor which enabled them to get started here in the first place.

Another introduced amphibian, the Marine Toad, I am happy to say has not been seen in recent years. This animal is a plague in some parts of Florida. Except for one or two initial specimens seen years ago, it has not been seen since.

The Narrowmouth Toad is about as abundant as it ever was. The Southern Toad is present and breeds sometimes in great abundance, and seems to be holding its own.

Chapter 25 - *RATS*

The two indigenous rats have been studied recently and are present in numbers in wild places, but in and around homes where the Cotton Rat used to be quite prevalent, it is being displaced by the alien Brown Rat, which is far too plentiful on Sanibel. They can be seen nightly in Cabbage Palms. People still tend to plant their Cabbage Palms too close to their houses, making it easy for the Brown Rat to enter roof spaces and other sheltered areas of the houses. The pitter-patter of little feet you hear in attic spaces all too frequently at night is usually the offensive Brown Rat.

The House Mouse is very plentiful, and on Sanibel it has taken the niche of wild species. The Cotton Mouse seems not to be present and perhaps has been displaced by the alien House Mouse. Some Cotton Mouse bones have been found in owl pellets on Sanibel. It is not clear whether these were from long flying owls that fed on the mainland, or whether the bones are of Sanibel origin.

Perusal of the following text will give the reader a good idea of what Sanibel has lost. Perhaps visitors and residents, new and old, may be induced by the changes here highlighted to take steps, before it is too late, to reverse the environmental deterioration that is marching apace on Sanibel today.

George R. Campbell
January 1, 1988

1 *THE RIVER OTTER* A Bright Light on Our Wildlife List

In April, 1975, on the "San-Cap Highway," the busy road between Sanibel and Captiva, a mother Otter was killed by a speeding motorist. The Otter had been lactating and thus, presumably, somewhere in the nearby wetlands there was a batch of pups slowly starving to death, vainly waiting for their mother's return.

A month earlier, an adult male Otter was crushed by a motorist on Periwinkle Way. On Easter Sunday, 1976, another male Otter met his death on the highway near the water plant.

People come to our islands to experience the tranquility of the wild places and to examine the wildlife, but the sad truth is that they come in such haste that much of our wildlife is being destroyed. The not too large population of Otters on Sanibel cannot stand this onslaught and it is hoped that readers of this book will join me and other conservationists in an effort

"In many ways man continues to be not only wild but more dangerous than any of the so-called wild animals". — Jawaharlal Nehru

to protect our wildlife by slower and more careful driving and by persuasion of others to do likewise.

Many naturalists have been pleased with the progress of the Otter on Sanibel and have determined that its relative success has been one of those fortunate but unintentional by-products of another activity, mosquito control. The mosquito ditches dug all over Sanibel provide a quite useful, although not exactly natural, habitat for the Otter. As a consequence, one can see Otters in many parts of the Island. Early in 1976 I saw a large one humping across the road near the "Panama" Canal. Dozens of times I have noted them out in the "Ding" Darling Refuge, both on the "fresh" and the salt water sides of the wildlife drive.

A party of young Otters was observed recently heading with their mother toward Jim Robson's Three Star Grocery. I'm not sure what they had in mind. Another was seen on Gulf Drive near the Bailey Tract. Behind the County "Park," near the water tank, there is a family of Otters living among the rubbish that the County has, characteristically, allowed to accumulate.

In short, there are quite a number of Otters around, but how sad it is for a whole litter of them to be orphaned and slowly starve to death in some wild spot on our "sanctuary" Island.

Otters are among the brightest lights in our wildlife list. They are truly playful creatures, with a lot of personality. They have special areas where they play and slide and wallow in the mud, much as school children do in their playgrounds.

Otters make fine pets and are among the most interesting and lovable of all wild animals. So if you ever do have to raise an orphan, it's not an unpleasant chore, although believe me, there is plenty of work involved. And for all their apparent bright friendliness, even the tamest of hand-raised Otters can turn on you in a moment of stress and give you a crushingly painful bite.

We have a resident on Sanibel, Emily Shane, of Gulf Drive, who is truly the Doctor Dolittle of the Otter. Emily has found them in all habitats, including the Gulf. She has photographed dozens of them and she seems somehow to have an "open wire" to the Otter brain and often succeeds in communicating with them. They seem to trust her, and she is able to follow them all about Sanibel. On one occasion she told me of an Otter that was lost in the Gulf, offshore from the main Gulf Beach. She talked to it and persuaded it to come ashore. She then helped it find its way back to the Sanibel River and saw it on its way into safer, more appropriate habitat.

Early in the morning, one can often see Emily Shane communicating with her Otters. A principal place where they might be seen is the turn on the "Ding" Darling Memorial Drive just past the Bird Tower. This is what I call Otter Corner. There, slides can be seen where the animals go up and down the banks on both sides. Usually the characteristic track, which has at least some tail-drag marking, is visible at this, a good place to view Otters.

In the summer of 1976 I watched an Otter cross the road from the Mangrove forest at Otter Corner, enter the borrow ditch on the "fresh-water" side, swim the ditch and walk out across the then exposed impoundment where it found a rather large stranded fish at about three hundred yards distance. The fish was also of interest to two Turkey Vultures and there followed a rather unusual confrontation between the Otter and the birds, during which the Otter seemed more intent upon trying to drive away the Vultures than doing what would seem to be the simplest thing — pick up the fish and take off for distant hidden parts. This period of "Otter confusion" lasted for the better part of an hour while we watched the drama through field glasses. The Otter finally managed to keep his fish when the Vultures became discouraged and left.

Our local Otter, called Common or River Otter, is *Lutra canadensis vaga.* There are several major Otter groups: the *Lutras* of North America and Eurasia; the Smooth Otters of Asia which are domesticated in Southeast Asia for fishing; the Clawless Otters of India; two rather strange creatures of Africa called Small Clawed Otters, and finally two very outstanding species, the Saro of Brazil and the Sea Otter of our own Pacific coast, both of which we will discuss briefly.

Of the Common Otters, there are twelve species, ranging all over South and North America as well as Eurasia. Our own subspecies, *vaga,* ranges throughout Florida, maritime Georgia and into a bit of coastal South Carolina. It is probably dependent upon fresh water but is often seen in marine situations also.

Two Sanibel Otters

The skull of the Otter is flattened dorsally and is equipped with unbelievable muscular strength. This flattened snout can be inserted into a space between two movable bodies, even very heavy ones, and can force them apart. A cage built to hold Otters must be very strong indeed. They can actually force their way through heavy steel wire mesh.

Otters are quite at home in the water and gracefully out-swim any possible enemy, including the Alligator. Their food consists of snails, crabs, shrimp, frogs, snakes, turtles, some water birds, rats and fish. The stomach contents of the killed female described above consisted of coot feathers and fish. There were a couple of pounds of other materials that I could not positively identify.

The Otter must eat an immense amount of highly proteinaceous food to support his physique and high energy lifestyle. An Otter will consume a quarter of its own weight every day, and this is appreciable when one considers that a big male might weigh thirty-five pounds. Nine pounds of flesh represents a lot of small animals. It is a good thing our Island habitats are sufficiently bio-productive to be able to support a population of such ravenous consumers.

Otters can stay a long time under water without surfacing, and when they finally do surface, the breath can be very surreptitious, only the flattened nose protruding for an instant before the animal disappears again under a "ring of bright water" for another long period. They are very strong and agile swimmers and spend a great deal of time patrolling a very large range on their hunting forays.

One of the most remarkable animals in the world is the Saro, *Pteronura* (some authors spell it *Pteroneura*) *brasiliensis,* otherwise known as the Giant River Otter. It is absolutely huge, as Otters go, attaining a length of seven feet! It is far and away the largest Otter in the world.

I spent many years of my life in South America and recall seeing this animal, before it became seriously endangered, in the Essequibo River region of Northern South America. Today it cannot be found in the abundance of the past because, like mammals everywhere, it has been persecuted for the value of its fur.

One can hardly blame a Brazilian *caboclo* for doing in one of these great animals, for he can sell a carefully-prepared pelt for the equivalent of five years' normal peasant income. And so it is that throughout the range of this animal, it has been over-hunted and is on the inexorable road to extinction. Recognizing this serious condition, the International Union for the Conservation of Nature (IUCN), headquartered in Morges, Switzerland, has placed this animal on "Red Book" status; that is, on its official endangered species list. The *Red Data Book* list is usually the guide followed by our own United States Department of Interior, and consequently the Saro is now officially classified by us as "Endangered." This simply means that its fur may no longer be legally imported into the United States. Thus half of the potential market has been wiped out. But France and West Germany import large numbers of the remaining stocks so that unless some more effective measures can be taken, this great animal will go the way of the Dodo.

Another fascinating Otter is the Sea Otter, *Enhydra lutris,* which has had some considerable influence on history, since it was responsible for the settling of our forty-ninth state, Alaska. This state was colonized by the Russians for no other reason than to secure Sea Otter pelts, which are equipped with very fine short but dense underwool and a longer overcoat of hair, both of which entrap insulating air. Unlike many marine mammals, the Sea Otter has no layer of subcutaneous fat. For hundreds of years, the Sea Otter pelt has been the most expensive of all furs and was used by Russian royalty for many generations. Catherine the Great would seldom wear the renowned Russian Sable, her furs being exclusively Sea Otter from Russian Alaska.

The species ranged from the Bering Straits down both sides of the Pacific to California on the east and to the Kuriles on the west. But over-hunting on the part of the Russians caused its depletion on the Asian coast, hence sparking their interest in Alaska, where the animal was soon over-exploited, almost to extinction.

In the American range, two subspecies are recognized: the Southern Sea Otter and the one from the Alaska coast. Although both races were hunted almost to death, both are making a modest comeback now.

The southern race faces another problem: the Abalone fishermen of California who are blaming the Otter for the shortage of their product. This is quite preposterous, for the blame really lies with the Abalone fishermen's greedy overutilization of this much-prized seafood resource.

Today there is a great gap in the former range along the Oregon, Washington and British Columbia coasts where no Sea Otters exist. There, industrial wastes are responsible for the Otter's inability to survive. We can also look forward to greater oil pollution in Alaska after the pipeline is operating to Prudhoe Bay, where even the most conservative thinkers feel sure that there will be spillage and concomitant destruction of marine resources.

The Sea Otter is unusual in another regard since it is one of those few wild creatures which, together with the Green Heron, the Chimpanzee and the Egyptian Vulture, are in fact true tool users. The Otter dives for mollusks amidst the great kelp beds off California and when it brings one to the surface, it also brings along a rock. The Otter then lies on its back with the rock on its chest and hammers the mollusk against this "anvil" until the soft parts of the mollusk are available for consumption. This sounds like a rather simple operation, but zoologists were really impressed when this behavior was first authenticated. In addition to mollusks, the Otter feeds on crabs, urchins and fish.

So you see the otters are really a fine group of animals and we on Sanibel are privileged to have our own beautiful race here. So please drive carefully and ask your friends to do likewise. Let's all brake when a wild thing crosses the road in front of us. Let's try to keep Sanibel a true wildlife sanctuary for all time.

Captiva

2 *THE ISLAND'S ONLY MARSUPIAL*
Johnny Come Lately

The Opossum, *Didelphis virginiana pigra,* is the only marsupial found on Sanibel Island. It is now well established here but only made its appearance after the bridge from the mainland was completed in 1963.

This animal is first mentioned in the "Ding" Darling Wildlife Refuge reports of 1963: "Several reports of Opossum sightings have been reported to personnel over the past months. Until completion of the Causeway, the Refuge lacked this mammal. Now the possibility of one or several reaching the Island by the new link and establishing the species here is to be expected."

That prophecy was certainly accurate, for the Opossum can be seen almost daily. Unfortunately, these sightings are all too often of animals smashed on the highway by motorists on our Sanctuary Island. It's a good thing Opossums are rather prolific and able to produce many young or they wouldn't long withstand the onslaught.

The Opossum belongs to the great and primitive order of mammals called the Marsupialia. The marsupials have reached their greatest development in North and South America and in Australia, where the only living representatives exist today. The group originated in North America, according to the fossil record, and reached South America when the Isthmus of Panama rose up from the sea. Those of Australasia arrived from South America via Antarctica when those continents were fused as one and were warm and forest covered. According to the now generally accepted theory of plate tectonics, Australia migrated from Antarctica about fifty or so million years ago and became the "Isolated Continent" as mammalogists call it, where largely non-placental or pouched animals evolved into their greatest variety and specialization—which heyday continues even to the present time.

Marsupials fit into almost all of the ecological niches that have been exploited by their placental relatives. There are subterranean burrowing marsupials rather like moles; there are also anteaters, not to mention counterparts of cats, bears, wolves, squirrels, rodents, horses and even deer.

Some Australian Marsupial forms actually "fly" through the air in the manner of our North American Flying Squirrels, *Glaucomys volans.* Some have a fully developed pouch or marsupium, where the young are nourished. Others simply have a couple of vestigial skin folds. Some are carnivorous, as is the almost (or possibly completely) extinct Tasmanian Wolf. Still others are purely herbivorous, as is the cute looking little "teddy bear" Koala that was saved from extinction by our President Herbert Hoover. It was Hoover who responded to the appeals of conservationists and prohibited the importation of the Koala's fur into the United States, thus saving that intriguing species for posterity.

I have some Australian flying marsupials called Sugar Gliders *(Petarus* sp.) in my collection on Sanibel. They are breeding quite successfully but are the most ill-natured little creatures that you can imagine. They use horrible language when anyone enters their room, but despite their temperament they are quite beautiful furry creatures with a flight membrane stretching on each side from the front leg to the back one.

Recently a zoologist from Queensland, Australia, was here and when he saw my little Sugar Gliders, asked about their diet. When I told him that they feed on a mixture of dogfood and bananas, plus chicken bones and so forth, with pediatric vitamin drops mixed in, he was a bit nonplussed and told me that I was feeding them a rather unnatural diet. They seem to be doing well on my diet but he suggested that I try giving them some *Melaleuca* flowers. This I did and these little creatures were delighted. My friend informed me that this is one of their principal dietary items in Australia and that it is the Sugar Glider that helps keep the *Melaleuca* in check in its native land. This of course suggests a similar course of action in South Florida where *Melaleuca* is completely out of hand and where, in some places, it's impossible to walk between them because they grow so close together.

The family Didelphidae, the only marsupial family that survives in the Americas, is very extensive. Its name derives from the fact that there are two wombs. The many representatives range from Canada to Argentina and *Didelphis* ranges throughout Eastern and Southern United States. Although it is not indigenous to our West Coast regions, the animal has been introduced there and is successful in California, Oregon, Washington and parts of Arizona and New Mexico.

Some three dozen marsupial taxa exist in the Americas, far fewer than are found in Australia. The only one that lives here with us is the above mentioned *Didelphis virginiana pigra,* and it is indeed a fascinating creature. Females give birth to up to eighteen young at one time, but since there are only thirteen "feeding stations," many don't make it. Commonly no more than six survive to reach an age where they are independent of the parents.

The gestation period of this animal is one of the shortest of all the mammals, something less than two weeks, and the tiny young are born in an extremely undeveloped condition. At birth the offspring is expelled through the urogenital sinus onto the base of the tail. From there it travels by itself to the pouch entrance by wriggling and scrambling along a path of fur dampened by the mother's tongue, using rather well developed, but deciduous, claws on the front feet. How such an immature, indeed embryonic, newborn can succeed in finding its way to the pouch is almost a miracle. When the little animal reaches the pouch, it firmly attaches itself to a teat which swells up in its mouth so that it is more or less permanently attached, at least for quite a number of weeks. As to whether the young marsupial can actually nurse or whether the mother forces milk into its mouth by means of muscular contractions is not known. The latter seems more reasonable to me considering the extremely undeveloped state of the newborn animal.

Photo, Laura Riley

Can you imagine ten baby Opossums in a pouch performing all of the bodily functions of baby Opossums in an almost completely enclosed space? Can you imagine what the atmosphere is like? Well, I can tell you this about it. There is from fifteen to twenty times the normal atmospheric concentration of carbon dioxide in there. One wonders how these animals are able to sustain life in such a vastly altered atmosphere. Also, I can well imagine what the smell would be like, but that doesn't seem to matter to infant Opossums.

Our Opossum is interesting in that it has hands and feet with opposable digits. Its big toe looks like an overgrown thumb. Another interesting appendage is its naked scaly ratty-looking tail that is prehensile so that this animal can actually hang from a branch by its tail, in the same manner as many of the South American monkeys.

The fur of the Opossum is used extensively as a cheap trim in American fashion, but in spite of this hunting pressure — plus the fact that rural people all over America eat its flesh — the Opossum is spreading successfully. It is succeeding where other animals are failing and can even be found today in most of the big cities of the North. I recall seeing Opossums in downtown Detroit on more than one occasion. One can only admire an animal that is so successful in a hostile world.

Of course, their omnivorous diet helps a lot — they can eat almost any organic matter that comes their way, such as insects, frogs, snakes, leaves, fruits, seeds, tubers and even carrion and garbage.

The old phrase "playing possum" comes from the habit of feigning death when disturbed or knocked about by a dog or other predator. An Opossum will lie without moving, waiting for the moment when escape may be possible after the predator is convinced that death has taken place.

This phenomenon is not confined to the Opossum, but has been observed in birds, reptiles and insects. The well known Hognose Snakes, of the genus *Heterodon* (one of which inhabits the mainland not far from Sanibel) display this same habit. When disturbed, it will swell up with air, loll its tongue out of its open mouth, and roll over on its back as though dead. Place it right side up, and over it goes again, thus giving away its deception.

Charles LeBuff witnessed an Opossum that had been annoyed by someone to the extent that it feigned death. When the opportunity offered itself, the animal recovered, dived into a canal and *walked* across the *bottom* to the other side, a distance of some fifteen feet, in a manner reminiscent of the Armadillo.

3 *THE FLORIDA PANTHER* And Other Sanibel Felines

Would you believe that even with today's very considerable environmental stresses the occasional Florida Panther is seen on Sanibel Island? After the Jaguar, the greatest cat of the Americas is the Panther — sometimes known as Cougar, Puma, Mountain Lion and Tigre. It can still be observed here one or two times each year. Of the fifteen or so races, the Florida Panther and the Eastern Cougar are under the greatest pressures, and are the most beleaguered and least likely to survive.

Once upon a time some over-enthusiastic mammalogist named more than one hundred and fifty races, or subspecies, of the Panther. But this sort of careless abandon in taxonomy has been rejected and today there are only about 15 recognized forms. In many ways, our own Florida Panther, *Felis concolor coryi,* is the most interesting.

The Panther ranges over the greatest area of any New World feline. It is found from Canada to Patagonia and from coast to coast on both American continents. This animal has been persecuted throughout most of the last two hundred years and its populations are vastly diminished in all areas. There is a widespread misconception that the panther is a great killer of domestic animals (a western race is even scientifically named *hippolestes* — horse killer) and this belief has been so well ingrained in the minds of ranchers and stock men that Panthers have been killed on sight and have been hunted ruthlessly through most of history since Columbus.

One of the most famous, some would say infamous, slayers of Panthers in Florida was one aptly-named Steve Hunter. He killed twenty-eight of the big cats before he himself was killed in a train wreck in 1963. In 1939, a few miles from Sanibel at Estero, a relative of Hunter killed a Panther that weighed slightly less than two hundred pounds. The all-time record was one killed at Eau Gallie in 1873; it weighed more than two hundred and forty pounds and measured nine feet, four inches from the tip of its nose to the end of its tail. More recently these animals have been killed near Bonita Springs and in the Big Cypress Swamp, south of Corkscrew Swamp Sanctuary.

In spite of its having been persecuted because of its alleged stock-killing activities ever since European man came to the Americas, modern mammalogists believe the Panther to prey primarily on smaller animals rather than on domestic animals, although occasional swine, colts and calves are no doubt taken. The major part of the diet, however, consists of rodents, Rabbits, Opossums and — especially here in Florida — Raccoons. When the Florida Panther occurred in normal numbers, its predation on Raccoons was influential in keeping their populations healthy and in check.

Panthers are born any time during the year in Florida, after a gestation period of three months. Litters consist of two or three youngsters. The young are spotted, unlike the adults which are generally of one color (hence the specific name *concolor*) and the tails of the youngsters are ringed. The distinct juvenile pattern gradually develops into the solid color of the adult animal, which in the case of the Florida Panther is rather different from that of other races — almost a dark gray with some rust and black hairs.

Other Panther subspecies have the more solid coloration, somewhat like that of the African Lion, and are substantially lighter in color than the Florida Panther. The color of the Florida Panther apparently varies in different parts of its range, with that of the Southwest Florida coastal region being considerably darker than those further inland.

What is the status of the Florida Panther today? Various authorities differ. The most recent estimate that I have seen is by the World Wildlife Fund, which states that there may be no more than ten to twenty Florida Panthers in South Florida. This differs from the International Union for the Conservation of Nature which has estimated that between one hundred fifty and three hundred exist and that they are actually spreading northward from South Florida into parts of their former range, even into Georgia.

The Florida Panther is an animal that requires rigid protection, which it has been afforded since 1966 under the law. But with ever-increasing habitat destruction due to development and spreading agriculture in this region, one is concerned about the future prospects for this important predator.

The IUCN, which can be described as the prestigious "U.N. of wild animal conservation and reporting," has stated that there have been occasional sightings in the "Ding" Darling National Wildlife Refuge here on Sanibel.

On the other side of the coin, "Uncle Clarence" Rutland, who has lived here for many decades and who has made many valuable wildlife observation records over the years, has never seen a Panther on Sanibel and cannot confirm that they have ever been here.

Equally valued observers, Willis and Opal Combs, sighted a Panther on Wulfert Road in 1960. In the summer of '74, Joe Wightman, a native of these parts who claims he knows a Panther when he sees one, states that he saw a Panther in the Wulfert area. There are many people on Sanibel, including such authorities as Charles LeBuff, who believe that the Cougar sometimes exists here.

One way to identify the existence of the Panther, in addition to sighting or finding tracks, is through its voice. My old friend Ross Allen had heard the Florida Panther scream many times at Allen's captive collection at Silver Springs, but he had never seen it in the act of screaming. He wrote a description of the first time he ever actually viewed this animal screaming:

"The Panther stood with her neck held straight out and her mouth pointed toward the ground. She then opened her mouth to about half its possible gape and voiced a series of loud, grating shrieks. The scream of this Panther would be likened to a high rasping human voice loudly shrieking 'ouch,' the sound being prolonged for about three or four seconds and repeated from three to seven times in succession."*

The official 1967 "Ding" Darling Refuge Annual Narrative Report to the Federal Government has this to say:

"A most unusual wildlife observation took place in the Darling area during February 1967. Several bird watchers arrived at the Refuge office and related their most convincing story of the sighting of a Florida Panther on the dike. From their adequate description, there is certainly a basis for our acceptance of the report. Another report of the cat's presence was received in June when an employee of the Mosquito Control District observed an individual in the vicinity of Turner Beach.

"That there is a representative of the Panther on the Refuge is highly possible as sightings are made frequently by competent naturalists on the nearby mainland. Territorial ranges of these cats are quite large and may include all terrain within a thirty-mile diameter circle. These cats are capable swimmers and quite feasibly an individual could periodically move down the chain of offshore islands. Food would create no problem for recent studies have indicated that perhaps seventy-five percent of the Panther's diet consists of Raccoons, and an inexhaustible supply is present on the Refuge."

Another sighting was related to me by Mrs. Eula Rhodes, whose home is on the Sanibel-Captiva Highway, near the Baptist Church. She saw a female and two baby Panthers in her yard in 1968. She is a very believable observer who said that the big cat's tail "almost touched the ground and curled up so pretty." She heard them "holler at night" and they made a trail through her garden. On being questioned, she said, "I *know* I saw them."

Rick Kraft and the late Walter Smith saw a dark cat the size of a small Shepherd Dog, with a long tail, near the power station on the Sanibel-Captiva Road on October 12, 1974. Mr. Kraft states, "I know it was a Panther." I questioned him in detail and I believe he saw a Panther. Mr. Kraft is a commercial airline pilot; people in his business have to be competent observers.

On November 18, 1975, Marion Campbell and Signe Cornelinson, both avid birders active in Audubon, were walking a few feet past the foot bridge in the "Ding" Darling Sanctuary. "Lo and behold, about thirty feet ahead of us," Signe Cornelinson wrote, "an animal was crossing the road. A dog? A bear? *NO,* it had a long curved tail, sleek bodied, low slung, dark colored, smallish head. 'A Panther,' we both exclaimed. Ah me! It made our day!"

The most recent believable Florida Panther observation was on Cayo Costa, a barrier island a few miles north of Sanibel. Jim Jenkins, a widely known TV personality now living on Cayo Costa, reports seeing a Panther near his place on November 22, 1976.

Lester Piper of Bonita Springs has captive bred a number of Florida Panthers and it is his observation that they are excellent swimmers. On one occasion he took a specimen to a canal in the Everglades and noted that it readily took to water and swam with skill. All of this makes it seem to me that this wide-ranging animal is visiting Sanibel, Captiva, Upper Captiva, Cayo Costa and possibly Pine Island, and the mainland region around Punta Rassa and Gladiolus Drive, where specimens have been observed recently. I believe it swims the passes at slack tide.

So keep your eyes open when you're traveling around Sanibel, and see if you can have the thrill of sighting this fascinating and increasingly rare great feline which still somehow manages to exist here in spite of all the pressures created by our overdevelopment. That none has ever been hit by a car in this region is certainly a tribute to its canny nature and good sense.

Another native feline is the Florida Bobcat, *Felis* (sometimes *Lynx) rufus floridanus,* which is seen with considerably more frequency than the foregoing species. I sighted one in January 1976 when my son, David, a trained zoologist, was with me and he confirmed and verified the sighting. Caroline Beebe saw one in the early evening of October 28, 1974, at the West Rocks. She positively identified it and says, "I'm sure it was a Bobcat. I do know a Bobcat when I see one."

We see enough of them so that we believe the center of distribution on Sanibel is in the area along Island Inn Road in the Sanibel Gardens—Bailey Tract Region, and the Sanibel-Captiva Conservation Foundation wetlands

***Quoted by permission**

The Florida Bobcat is a case of conservation backfire.

area. They're not all that rare. We see cat tracks that are rather larger than those of the feral domestic cat. We know that the Bobcat is present in some numbers on the nearby mainland, where I saw one smashed on the road by Page Field in late 1976. I know a lady who keeps one in a cage near Miner's Plaza and who shot another one that came to visit. So although certainly under pressure and diminishing in numbers, the Florida Bobcat is still here.

Now that Leopard, Cheetah and Jaguar furs are illegal in the United States, Bobcats are being destroyed in ever-increasing numbers. The furs are usually marketed as "Lynx". In early 1977 Jackie Onassis was mentioned as wearing her new Lynx coat at an art show in New York where this fur, about the only legal "spotted" cat fur left, is becoming far too popular. It doesn't even make good durable fur since it is fragile and has brittle hairs, although that is not reflected in the prices being charged—up to $5,000 for a coat. Personally, I believe the only ones who should wear Bobcat coats are Bobcats!

This is a case of conservation backfire—the limited protection our Endangered Species Law gives to the Leopards has increased the pressures on the unprotected "varmint," our beautiful Bobcat. There is a movement in process to have it declared officially "endangered" by our government.

The third cat that occurs on Sanibel Island should not. I refer to the very large population of feral domestic cats, *Felis catus,* which is truly a menace, in more ways than one. For some reason, thoughtless people visit Sanibel Island and dump their unwanted cats and kittens onto our Sanctuary Island where they do untold harm to our wildlife, preying on small birds and mammals as well as lizards and snakes. They become scrawny and diseased and are certainly not desirable additions to our animal community. Many are not only sick themselves, but carry feline diseases that attack wild animals, especially the Raccoon.

Many misguided people feed these feral cats and this only serves to perpetuate a bad situation. Better humanely to trap and destroy them, or to adopt them. Many fine pet cats have been adopted from this feral population and certainly nobody can quarrel with that, but to feed and thus help to perpetuate the wild population, and to add to them by bringing in new animals, is certainly undesirable.

4 *DEMISE OF THE WHALES* Mankind's Great Folly

Do you know that there are about a dozen whale species that have been sighted in the waters around Sanibel? And there are probably that many others in the region for which records may or should exist, but which have not yet been definitely verified. A list of these as well as other Sanibel mammals will be found at the end of this book.

Probably the most exciting forms that have been sighted around here are the Atlantic Killer Whale, *Grampus orca,* (*Orcinus orca*) and the False Killer Whale, *Pseudorca crassidens;* the former is rarely seen, the latter sometimes appears in large pods.

Of course the one that is best known hereabouts is our own little Atlantic Bottlenose Dolphin, *Tursiops truncatus.* This is the species that was made famous by "Flipper" and is the one seen so frequently near the Sanibel Causeway.

It is very commonly seen in oceanaria around the world and is capable of learning intricate routines that can put a dog or an ape to shame. They are so intelligent that some cetologists feel that communication on or near our own human intellectual level is possible.

Many workers feel that some other cetacean species also approach the human animal in intellectual ability. Cetaceans are thought to be the brightest group of mammals in the world after the primates and some bold scientists even believe they are in some ways more intelligent than the primates. Intelligence, however, is very difficult to define and it is next to impossible to set intelligence criteria for whales.

The cetaceans are a large group of mammals that are entirely aquatic. There are two main groups today: the Odontoceti, or the toothed whales, a good example of which is the Bottlenose Dolphin, and the Mysticeti, or the Baleen Whales.

Instead of teeth, Baleen Whales are equipped with modified "gums" which act as strainers for the removal of their food which consists of the tiny suspended crustacea which, with other small animals and plants, are collectively called plankton.

The best known Baleen Whale, the Great Blue Whale, *Balaenoptera musculus,* is the largest animal that ever lived on this earth, larger even than the largest dinosaurs.

There are over ninety kinds of whales and dolphins in these two groups and they occur in all of the seas of the world and in some rivers and lakes. However, many forms occur in ever-decreasing numbers due to the greedy overkill on the part of protein hungry mankind. Today the Japanese and the Soviets are the main offenders, most other nations having ceased their pursuit of whales.

Let me say here that I think of all cetaceans as whales and in fact I believe that it is perfectly correct to refer to them by that name. If I call our little Bottlenose *Tursiops* a whale, please bear with me. And if you pick up that habit you too will probably find it more exciting to think of them as whales which they are—rather than simply porpoises. We might as well clear up another ambiguity of nomenclature while we are at it; I don't recognize any difference in meaning between "porpoises" and "dolphins" (except for the very beautiful tropical Dolphin Fish of the family Coryphaenidae, of course). As far as I am concerned, the words can be used interchangeably when discussing cetaceans and a lot of cetologists are in agreement.

Imagine being able to pick up a live whale and hold it in your arms, a live whale little larger than a dog. One of the few fresh-water whale species is the little Susu, *Platanista gangetica,* the little pointy-nosed dolphin of the Ganges and Indus Rivers of the Indian

Sub-Continent. These little animals go far upstream and were quite plentiful until recently. They are taken for oil and folk medical purposes by the local people. Their present status is to be the subject of an IUCN study.

The youngsters are of a size that can actually be held in your arms. This little animal has no sight. It is completely blind because it and its ancestors have been poking around the murky bottoms of two of the most turbid rivers in the world for many thousands of years. The waters of the Ganges and the Indus would offer no visual opportunity for any animal even if it had the keenest sight. The Susu probes for fish and crustacea with its sensitive snout.

As warm blooded animals, whales must breathe at the surface. The old whaling term "thar she blows" refers to a whale's blasting forth the contents of its lungs at the surface of the water just prior to taking a new breath. Contrary to popular belief, whales do not blow liquid water out of their lungs; the visible spout is simply the condensation of water vapor in the warm air from their lungs and possibly the discharge of mucus foam which fills the air sinuses of the respiratory system.

Before making a deep dive, the whale expels the air from its lungs. This seems unusual considering how long whales can stay under and how deep they can go. It is made possible by the fact that the oxygen from the air breathed in at the surface is promptly combined with the hemoglobin of the blood and the myoglobin of the muscles.

When we humans dive, we carry with us an air supply in our lungs, but since the whale has already utilized his oxygen supply and built it into his blood and muscle tissue before he dives, he can dive on relatively empty lungs, thus lowering his buoyancy and permitting deeper dives with the expenditure of less energy.

Whales have a neat switching system for blood that shunts aerated blood to the brain and muscle on a deep dive. Other areas which are not working, say the digestive tract or the reproductive organs or perhaps the lungs themselves, may thus be cut off from the large amounts of oxygenated blood at these times when they don't need it. This permits the brain and other essential areas that are being utilized during the dive to get the lion's share (or should we say "get a whale of a lot") of oxygenated blood. During a deep dive, the whale's heartbeat is decreased and the respiratory center of the brain is insensitive to the accumulation of carbon dioxide in the blood and tissues which triggers normal breathing reflexes in most mammals.

Deep dives are safer for a whale than they are for you or me because the whale is not subject to "the bends." When we go down, nitrogen and other gases of the air diffuse through our lungs into our blood. Sudden surfacing (i.e., suddenly reduced pressure) causes the bubbly bends that are so painful and dangerous to human divers. But the whale does not take down lungs full of gas and has the oxygen already built into his tissues. Consequently this problem is avoided.

The Sperm Whale (the Moby Dick Type), *Physeter catodon,* was the preferred target of our historic New England whaling industry in earlier centuries. They are still hunted today although there are probably fewer than one hundred and fifty thousand of them left. In 1930 there were six hundred thousand; before intensive mechanized hunting began in the last century, there were probably more than twice that many.

It has been said that the Sperm Whale's brain is probably the most marvelously intricate biological entity ever to have evolved in nature. Yet this wonder of nature is still stuffed into cans for dog food by some of the "civilized" nations of the world. To the credit of the United States, this practice no longer takes place in our country.

There is no longer a whaling industry of any kind in this country and all of the great whales are included on the U.S. official endangered species list. No parts, no meat, no oil, not even scrimshaw may be brought into this country and entered into commercial transactions. Scrimshaw, as you may know, is the artwork that was carved on whale ivory, usually Sperm Whale teeth, by bored seamen suffering from the ennui of long, perhaps three year, voyages.

There is one Sperm Whale product which may be legally permitted, or at least should be in some cases—that is, when found washed up on the beach. I refer to ambergris. Ambergris is a waxy, smelly substance that comes from the Sperm Whale's gastrointestinal tract. It was very useful in the perfume trade as a scent stabilizer. My old marine biology professor, F. G. Walton Smith, founder of Planet Ocean, colorfully described ambergris as a "billiary concretion of a sick Sperm Whale." As such, it doesn't sound very appetizing, and truly, it isn't.

I recall how one day some 40 years ago, I stumbled on a blob of over one hundred pounds of ambergris at Palm Beach. I recognized it, hid the bulk of it high above the tides in dune vegetation, whacked off a foul smelling sample and sent it off to a company in New York who were famous dealers in strange natural substances. I was informed in reply that ambergris had fallen into disuse by the perfume industry and was no longer used extensively, but otherwise mine was perfectly good. Had it been my grandfather who stumbled onto this treasure, mine could have become a wealthy family. Ambergris used to sell for large sums, its value being greater than that of gold. It was just my luck to find this substance after it had been phased out of the market. (I understand that now it has regained some of its old value and is again being sought.)

Once, when I lived on the island of Eleuthera, a Sperm Whale beached itself and died on the Atlantic shore at the edge of the Continental Shelf. The whole region virtually closed down. As the tons of flesh began to rot, the stench could be detected for ten miles if the wind was right. Eleuthera, being a tourist island, found the visitors moving out in droves, which occurrence did have its compensatory value: our summer peace came earlier that year.

What do you suppose would happen if a large whale committed suicide on Sanibel's fabled shell beach down near the condominium concentration during the height of the season?

Let's just say a word about the Blue Whale before we go on to the fate of the whales and what should be done about them. The great Blue Whale as mentioned earlier is the largest animal that ever existed; it can weigh four hundred thirty thousand pounds and can be up to one hundred feet long. One of my associates, L. A. McHargue, who worked valiantly on whale conservation issues, changed that around a little bit and said "The Great Blue Whale is the largest animal ever to die." For death has been the fate of the Blue Whale. It has been driven almost to the edge of extinction. There may be no more than six hundred of them left today. As recently as 1930 there were one hundred thousand and before the great push on commercial whaling started in the last century, there were probably ten times that many.

The question today is whether there are enough of them left to find each other in the vast expanses of the oceans so that reproduction can take place. After all, the Atlantic, Pacific and Antarctic seas are immense and 500 or 600 whales might have trouble finding each other. They do not breed when they are suckling their young and at least two years pass between successive pregnancies. The gestation period is almost a full year. Twins are rare, single births being the common occurrence. So how are so few to reproduce themselves in a hostile world where this animal is still taken by the Japanese and the Russians and by some shore stations of other countries?

Since early this century when intensive hunting of whales began, the reduction has been immense. The combined population of the eight species of great whales (the Atlantic Grey, the Southern Right, the Blue, the Hump-backed, the Pacific Grey, the Fin, the Sei and the Sperm Whale) in the early days of this century totalled an estimated four and a half million. By 1930 they had been reduced to 1.6 million and today there are no more than two hundred fifty thousand of these great whales left in all of the seas of the world.

The tragedy of the demise of the great whales certainly represents one of mankind's greatest follies.

For many years, an organization known as the International Whaling Commission has been meeting each year, mostly in Washington or London. The delegates are sent by the whaling nations plus the U.S.A. and Britain. Today some nations such as Brazil, Argentina, South Africa and Norway are represented, nations that used to conduct pelagic whaling operations, but no longer do. However, some still do operate shore stations.

The International Whaling Commission over the years gained a reputation that is hardly to be envied. They are noted for their do-nothing, soft line attitude toward the preservation of the great whales. Until a year or two ago they accomplished very little. Accordingly, some call this body the International Whaling Omission. Their resolutions are not bound by law and are simply taken home to each nation where they are often ignored, especially when they cut into the profits of the whaling industry.

Public opinion against the slaughter of the great whales has been gaining strength, however, and some progress has been made. The U.S.A. until quite recently had a shore whaling station in California, but this is no longer in existence. Canada had a station in the Maritimes, which has been closed down. South Africa, Brazil and Argentina still have shore stations, One in Brazil is owned and operated by the Japanese.

Shore station whaling, as contrasted with pelagic whaling, is not considered to be as destructive, since many fewer animals are killed with this method. Small boats go to sea, harpoon and float dead whales to shore for flensing and preparation.

It is the pelagic whaling of the Japanese and Soviets that is so very, very destructive. Pelagic whaling employs huge factory ships with stern slide gates, crewed with hundreds of specialist workers, equipped with rendering vats, boilers and refrigerators. The slide gates permit the loading of even the largest whale onto the preparation deck.

In a very short while, fifty tons of whale can be reduced to oil, flesh and other whale products and another leviathan carcass can be brought on board. Each factory ship has crowded about it a fleet of smaller vessels, as a hen has her chicks. These go out to seek the pods of whales which are often located by helicopter or by other sophisticated devices. Electronic devices of modern warfare are employed against these huge defenseless and highly intelligent animals.

When a hunting vessel comes upon a pod, the destruction is immense and immediate and the sea for a mile around is stained by blood. Harpoon guns are equipped with explosive devices which are thrust into the body cavities of whales. The explosion takes place in the viscera and often a slow death results.

This whole operation can hardly be endorsed by any decent society. In fact, at the International Conference on the Environment in Stockholm in 1972, the major theme called for a ten-year moratorium on the killing of all great whales. This moratorium was resolved by a majority of delegates at "Stockholm '72' but the Japanese and the Soviets continue to ignore world opinion and pelagic whaling continues at a devastating pace by these two nations.

The Soviets don't offer the world any reason for their action; in fact, they hardly comment at all. The Japanese, however, are sensitive to the condition of their world market and they constantly make feeble excuses for their continued disregard for international opinion on the whaling moratorium.

In the last couple years the Fund for Animals and the Audubon Societies of America, together with other conservation groups, declared a boycott of Japanese and Soviet goods until the unconscionable slaughter of the great whales is stopped. These actions have had a telling effect on the Japanese, and their pelagic whaling industry is definitely on the decline. There is reason to believe that it will cease in the next two or three years. Perhaps the U.S.S.R. will then follow suit. Whether this will be in time to save some of the great whale species is not known, but there are grave doubts.

Another whale problem deals with the small species of the Pacific. The Yellow-finned Tuna is located by means of porpoise activity. Yellow-finned Tuna exist in great schools at some depth in the waters of the Pacific off central and South America. For some reason that is not clearly understood, great schools of porpoises swim above the tuna and travel with them. The U.S. tuna fleets which are based in San Diego, and also those of Japan, use this natural phenomenon to locate the tuna.

A relatively new U.S. invention — the purse seine — a huge net, shaped like a purse, is laid completely around the schools of both porpoises and tuna and slowly hauled into a smaller and smaller area. As a result of entanglement in the net during this process, somewhere around a quarter of a million porpoises are drowned each year by the U.S. fleet. In some past years the mortality was believed to be as high as nine hundred thousand.

This the tuna industry excuses as "incidental" kill. Since it is not doing this intentionally and in fact does not even save the drowned cetaceans, but simply wastes them overboard, the industry holds itself blameless.

A master of a San Diego tuna boat, one Captain Medina, developed what is called the Medina Panel, which is a special net hanging at the surface of the purse seine that can be manipulated to permit most of the porpoises, but not the tuna, to escape. This device is gaining some acceptance, but the tuna lobby in Washington is strong and there is leniency toward the tuna industry in spite of the Marine Mammal Protection Act of 1972 which man-

dated the end of "incidental" kill. Recent court action has given some support to the law, but the eventual outcome is not predictable.

The Japanese kill is also very large — perhaps two hundred thousand or more per year. There is even less hope for Japan for there is evidence that the Japanese are no longer considering this kill to be "incidental." Rather, with the diminishing stocks of the great whales, the Japanese are thought to be seeking out these smaller ones in order to compensate for the losses experienced by their whaling fleets.

There is a decent alternative to pelagic whaling. The Norwegians, who were great pelagic whalers in the past, have converted their factory ships into fish protein concentrate factories (FPC). In some cases, not fish but crustaceans are utilized (CPC).

Some of the recently-exploited huge concentrations of fish have declined in quantity in recent years. For example, the billions of South American anchovy have been declining. Peru's economic boom of the sixties was based on overuse of its anchovy resource—overuse to the extent that the guano or bird dropping fertilizer industry was almost wiped out. The anchovy fleets worked six days and rested the seventh. "Only on Sunday" were the birds able to feed. By the seventies the resource was almost fished out and Peru was in a depression. Thus there are not always fish stocks available to factory ships. But one substitute for fish has increased due to the decline of the great whales, namely Krill and other planktonic organisms of the Antarctic Seas which constituted the mainstay diet of the Baleen Whales.

It takes a lot of plankton to feed a Blue Whale that might weigh 200,000 pounds, and since these whales have been greatly reduced in numbers, down to perhaps 600, it is obvious that there could be a superabundance of left-over whale food in the southern seas.

One ecologist of my acquaintance is concerned about this "overbloom" of Krill and other organisms in the Antarctic Seas. I noted recently that Cousteau also remarked on this phenomenon. So conversion of factory ships from whaling to plankton utilization, as has already been pioneered by some Scandinavians, just may be a useful road to follow in this protein-starved, overpopulated world that we live in. Of course, if this resulted in a recovery of great whale populations, a potential problem of humans and whales competing for the same resource is inevitable.

History is replete with stories of how various whales at one time or another lent a helping hand to man. The ancient Greeks cited many legends of porpoises saving drowning seamen.

One of the most unbelievable, but true, events relates to a pod of Killer Whales that, for two generations, carried on a symbiotic relationship with fishermen in southern Australia, at a fishing village called Eden. This is well documented in a book called *Killers of Eden*[1] which tells of how the fishermen would go to sea, signal to their pod of Killer Whales and in a cooperative effort, man and whale would fish. The whales drove the schools of fish into the nets which were hauled in by the men after which there was an equitable sharing of the catch.

A similar, but not nearly so well documented, relationship between man and whale took place on the southern coast of Brazil early this century. And currently, there is a well authenticated report of human-whale cooperation on the Atlantic coast of Mauritania in northwest Africa, where a native fishing population of tribal Imraguens is aided by a local cetacean pod. Each year, between September and April, huge numbers of mullet migrate past their shores. Each morning, during that period, the fishermen scan the sea for the slight color change that tells them that the great schools of mullet are present. Such a school may be a quarter of a mile long, many yards wide and six feet thick. When a school is detected, a fisherman spanks the water surface briskly with a boat paddle, producing a sound that carries underwater for long distances. Many dolphins respond and begin to appear almost immediately on the horizon. They drive the fish into the nets held by teams of Imraguens. The dolphins block the net entrances, consume their fully earned share as their "commission" and depart, while the men haul their tons of fish to shore for sun-drying and storage. This symbiotic relationship benefits both man and dolphin. Only the "poor fish" is the loser.

The whale species that is best known to us here on Sanibel and the one that has had the greatest living experience (there are many that have had dying experience) with mankind, is our own little *Tursiops truncatus,* the Bottlenose Dolphin, which is so commonly trained in many marine aquaria all over the world.

Until the Marine Mammal Protection Act came into force a few years ago, there was quite an active business in the Florida Keys of capturing porpoises and shipping them off by air to institutions throughout the world. This activity has now been greatly curtailed. Any porpoises taken today by U.S. citizens must be taken under Federal government permit, even if taken outside of U.S.-owned or controlled waters.

Most wild animals, even forms very high on the phylogenetic tree such as the Chimpanzee, man's closest relative, must be badly mistreated to be trained for show business. Chimpanzees are regularly beaten to insure their cooperation in intricate showtime routines, certainly an indefensible practice. I am personally acquainted, for example, with the torture, maiming and even killing of Chimpanzees that took place at the Detroit Zoo in order to produce a split second, intricately performed chimp show.

This type of training is not necessary with our little *Tursiops.* It's a cooperative intellectual being which soon acquires rapport with its human associates. It is eager to cooperate and eager to perform whatever is asked of it. A remarkably intellectual relationship between trainer and animal develops.

Dr. John Lilly worked for many years trying to discover a common language possible to both man and dolphin but this has so far eluded him. Such work continues and every year new and fascinating aspects of cetacean behavior are revealed.

The late Captain William B. Gray, who for many years was the Director of Collections and Exhibitions at the Miami Seaquarium, told many interesting porpoise anecdotes over the years. It was Captain Gray who captured the

[1]*Killers of Eden,* Angus & Robertson, Sydney, 1961

This False Killer Whale beached himself and died.

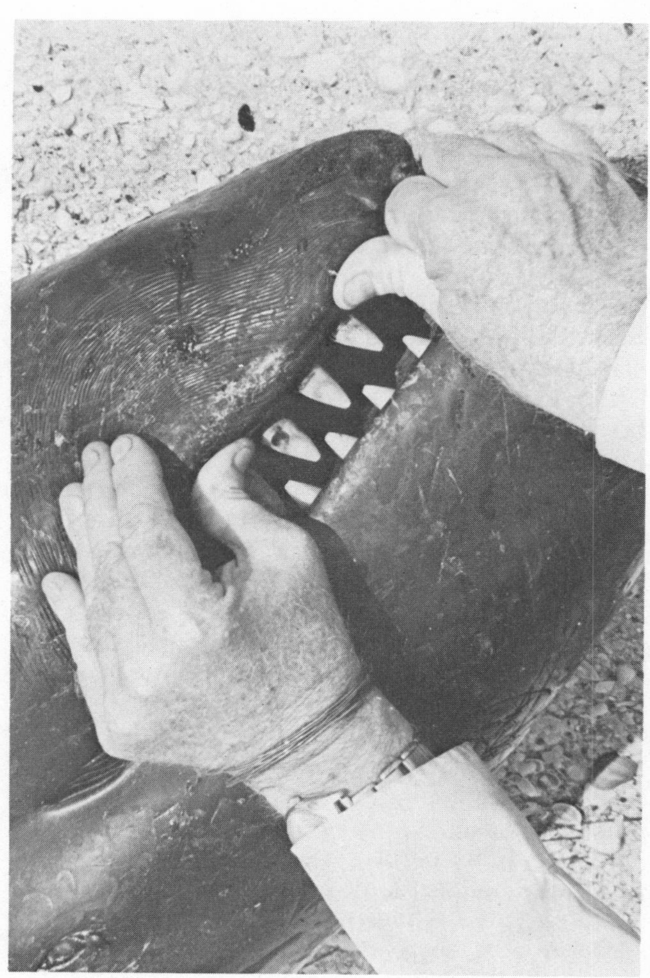

Photos, Peter Larson, *Island Reporter*

The teeth are quite formidable.

albino *Tursiops* that lived for several years in Miami and became known to millions of people as "Carolina Snowball". Unfortunately she didn't survive to breed and Captain Gray's hopes of developing a white race of porpoises were frustrated.

Gray used to capture most of his porpoises in shallow inlet water along the Intercoastal Waterway on the East Coast of Florida, using stop nets. The porpoises would come from the sea into the inland water after fish on a high tide. After gorging they would return on the ebb to the ocean. Gray would simply close off the inlet using two stop nets and would encircle the porpoise drawing him into an ever smaller bunt, or pocket, until the animal could be tied to a stretcher, kept damp with wet blankets and trucked to Miami. Interestingly enough, these highly intelligent wild animals would usually not try to jump three inches over the net's corkline although later in the Seaquarium the same animals could be trained to jump more than twenty feet into the air.

Porpoises are in motion at all times; even when they are asleep, they swim. They are very curious and will play with almost anything at all in the tank, whether it be a ball or a ring or a jet of water, or even a wad of chewing gum dropped by a careless visitor.

Porpoises solve problems very easily. I once watched five *Tursiops* in a tank organize themselves with about as much discipline as a football team in order to enable one of them to make a jump of some twenty feet to his trainer's hand for a morsel of fish. The performer lined up his four companions in a neat line out of the way where they stayed and watched the proceedings with excited intent, but nervously restrained themselves from any participation. The performer then circled his huge tank, nudging a Loggerhead Turtle out of the way with his nose, waiting for a Lemon Shark to pass and for a Sawfish which was obstructing the way to leave the chosen path. Then, after all was in perfect order, just like a boy trying for the highjump, he speedily made his prerun across the diameter of the bottom of the tank, accomplished a ninety-degree turn to the vertical and hit the surface at tremendous speed, jumping into the air for his fish reward. The organization and the planning he had perfected before his feat were more remarkable than the high jump itself, and certainly demonstrated a high level of reasoning and forethought.

Porpoises have a very strong sense of maternal responsibility, as one might expect in an animal of such high intelligence. Mother porpoises take excellent care of their young. As soon as a baby is born, the mother or any other female in the pod will push it to the surface so that it can take its first breath. Immediately thereafter the animal is ready to swim with the others although it will not be weaned for eighteen months.

Captain Gray once told me of a porpoise that was given a medication in a capsule hidden in a Bluefish. The porpoise ate the fish and the medicine did its work but that animal never touched another Bluefish for the rest of its life.

Some behavioral mammalogists compare the affectionate interaction in whales with that which is common and well known in the primates. It is possible that the long time span required for the development of sexual and physical maturity in both the primates and the cetaceans results in analogous patterns of learned behavior. For some members of both groups there is a familiar progression of natural social experiences beginning with infant-mother interactions, then progressing to youngsters rough-housing, and heterosexual relationships eventuating in reproduction. The mother receives valuable rewards to compensate her for the loss of freedom entailed in rearing youngsters over a long period. It is interesting and probably not coincidental that two such varied and unrelated groups would pass through similar developmental processes.

Dolphins, like young primates, play a great deal of the time. They originate complex games using almost any foriegn object available to them. Dolphins will even play games with human visitors at an oceanarium. A ball on the surface may be passed back and forth to a child who might by chance be visiting and not be known to the dolphin. Visitors at oceanaria sometimes throw coins to the dolphins in an attempt to simulate interaction. This can be fun, but dangerous, for the animals tend to swallow small foreign objects and there are records of resultant deaths. Swallowing such objects seems to be a result of the competitive nature existing between two or more of these animals; rather than allow another dolphin to take away its possession, the animal may swallow that object.

Many people condemn the keeping of cetaceans in captivity. Many are kept under horrible conditions — such as those I saw overcrowded and in polluted water at a Galveston, Texas, shopping center tank show that was supposed to stimulate business. Also, the military use of these animals is indefensible. But for well cared for animals in good oceanaria, I feel the beneficial results may be worth the sacrifice of a few animals. For today people all over the world have an interest in and sympathy for these creatures that has certainly been fostered by their having been exhibited and shown in mass media. Such a trade-off may be worth the price that is undoubtedly paid by the death of some.

In July of 1976 we experienced some tragic mass beachings of the False Killer Whale, *Pseudorca crassidens,* in this region. On the 16th of July a pod of about twenty was seen off the coast of Sanibel in shallow water, only a few feet from our beach. We couldn't tell for sure whether or not they were in trouble and shortly thereafter they moved out of sight toward the coast of our neighboring island, Captiva. This seemed to be unnatural behavior, for this species is considered to be a deep sea animal. A few days later I got word that they were trying to beach themselves on an island a bit further to the north. It was on that day that I learned that in an emergency it is possible to hitchhike by air and water throughout these parts. I was able to secure a flight on a passing

Photo, Peter Larson, *Island Reporter*

Peter Larson and the author herded the remaining False Killer Whales out to sea in an effort to keep them from running ashore. Our success was minimal, however, for we believe all later died some hundred miles south of Sanibel.

Photo, Peter Larson, *Island Reporter*

Many kind people spent long hours protecting the stranded False Killer Whales from the sun with wet towels and lanolin rubs.

helicopter generously supplied by Wayne Miller. We went over the area and saw the whales heading toward shore. I asked the pilot to set the chopper down on the beach. Peter Larson and I flagged down a passing cruiser, explained our interest in the whales to the owner and were invited on board. We then proceeded to head out to sea in an effort to turn the pod around and get the animals to leave the area and head for deep water. In this we were successful, but this same pod had earlier lost five of its members. One was dead at the north point of Captiva and four others were stranded on the inside of the islands on a bar in the Bay where for many hours various citizens took care of them in an effort to keep them alive until help could come from an oceanarium called Sea World, where facilities exist which might have saved the whales.

They were treated with lanolin to prevent sunburn and were kept damp; many people worked among them for long periods in an effort to save them. They were later trucked away to Sea World, but none survived for very long.

The pod that left for sea turned up a few days later at Loggerhead Key, near Dry Tortugas, where again they were encouraged to head for deep water, but a few weeks later many great carcasses were found beached in the Everglades National Park at the southern tip of the Florida peninsula.

This interesting experience was very tragic and depressing. Cetologists are still trying to find out why whales sometimes beach themselves. One theory is that parasitic infestations of the middle and inner ear may cause disequilibrium and disorientation in the animals. Or perhaps the infestation occurs only in the lead animal and the others merely follow. However, most mammals have a normal load of parasites that exist quite successfully without harming the host and this fact, of course, casts doubt on the theory.

Beachings are not unusual in Florida, there having been many recorded among several species in the past. Another recent occurrence took place in 1971 when more than 150 False Killer Whales were stranded on the Atlantic side of the peninsula. In another incident in February 1977, hundreds of Pilot Whales, *Globicephala macrorhyncha,* committed mass suicide near Jacksonville. Smithsonian marine mammalogist Dr. James Mead who was there directing necropsy activities was asked when science might find the answers to these mass beachings. "Twenty or thirty years?" was his reply. "All we're going to be able to do is guess. We're going to be dealing with theories for a long, long time."

I hope that this chapter will serve to elucidate some of the values and problems of the cetaceans and to demonstrate that this perhaps the most interesting group of animals in the world, is in dire straits. Many forms are facing extinction in the next few years unless some intelligent effective international action is taken now.

As long ago as the mid-1800's, when Herman Melville wrote his famous *Moby Dick,* there was concern for the future of the whales as will be attested from this quote from that book:

"But still another inquiry remains; one often agitated by the more recondite Nantucketers. Whether owing to the almost omniscient lookouts at the mastheads of the whale ships, now penetrating even through Behring's Straits, and into the remotest secret drawers and lockers of the world; and the thousand harpoons and lances darted along all continental coasts; the moot point is, whether Leviathan can long endure so wide a chase, and so remorseless a havoc; whether he must not at last be exterminated from the waters, and the last whale, like the last man, smoke his last pipe, and then himself evaporate in the final puff."

Some scientists have learned to identify individual whales by the different shapes of their dorsal fins. Famed elephant student Douglas-Hamilton does much the same with his subjects — but in his case, ear shape is employed.

Photo, Peter Larson, *Island Reporter*

5 *THE ANCIENT ALLIGATOR—MESOZOIC LEFTOVER*
A Plea for Detente

There is one animal on Sanibel Island that is treated with excessive kindness by some soft-hearted and well-meaning people and as Public Enemy Number One by others. Both groups are wrong. The "softy" school insists on feeding the beast—on fish scraps, bread and, incredibly, marshmallows. This should not be done. The other group would just as soon see the species completely wiped out, which is morally wrong and economically unwise.

The victim of all this misplaced affection and misguided venom is the American Alligator, one of Sanibel's most valuable animals.

Sanibel has a viable and substantial population of wild Alligators and these animals are very popular with our visitors. They are, in fact, the most sought after by wildlife watchers of all our abundant fauna species.

Of the two dozen or so crocodilians that remain alive on this much abused planet, the American Alligator is the most gentle and tractable. Comparing it, for example, with Asia's terrific engine of destruction, the great Salt Water Crocodile, *Crocodylus porosus,* is like comparing a Toy Poodle with a savage Doberman.

Yet our Alligator can be transformed into a dangerous animal by man's thoughtless "taming"—which usually means feeding. The principle here is exactly the same as that of the bear feeding problem in our National Parks. Indeed, it is so like that problem that I have dubbed it "the Yellowstone Bear Syndrome."

A "tame" fed Alligator, if left unfed for whatever reason, may become a potentially dangerous animal when he wanders into a yard, swimming pool or garage in search of food. Animals not thus fed are less likely to become problems. A fed Alligator becomes bold as he loses his fear of man. Some have even been fed by hand, and an Alligator does not really have the intellectual capacity to differentiate between a piece of fish and the hand that holds it.

Recognizing this problem, Sanibel City Council wisely passed an ordinance "prohibiting the feeding of American Alligators within the corporate limits of Sanibel." This ordinance was written by Councilman Charles LeBuff who, in addition to being active in civic affairs, the "Ding" Darling National Wildlife Refuge and the Loggerhead Sea Turtle preservation group known as Caretta Research, is also one of the region's leading herpetologists and has done much research on the Alligator.

Although the Alligator is the principal attraction of most tourists who visit the Island (a fact that some of our merchants and business people tend to overlook), the animal is hated by many who live here. Why this is true is difficult to explain. But a review of the natural history of the Alligator and the values that this species contributes to our sanctuary environment may shed some light on the situation.

Borrowing a term which I believe was originated by one of our resident ornithologists, George Weymouth, the Alligator "patrols the rookery." Insular rookeries attract Alligators. The reptiles seek young birds which may fall out of nests, as well as old and unhealthy birds, or any others they can grip in their powerful jaws. This activity is not harmful to the bird populations, but rather supports them, because the Alligators will not allow other predators—especially the tree-climbing ones—to enter. An unwise Raccoon who seeks to cross the water barrier to an insular rookery often winds up in the stomach of an Alligator.

Another well known and valuable habit of the Alligator is the digging of the "gator holes." A gator hole is simply a pond created by an Alligator and maintained with sufficient depth to provide year-round fresh water. To make a pond, an Alligator uproots grass and mud, slicing with his mighty tail, digging with his powerful feet, and carrying debris away in his mouth. The resulting pond is rather deeper than surrounding land and consequently as the drought progresses into the dry season, in some localities the only natural fresh water is in the Alligator hole. Such a bit of isolated fresh water allows for the maintenance of many, many other groups of animals including reptiles, fish, birds, amphibians, mammals and many invertebrates such as snails, shrimp and crayfish. When the rains come again there will be a dispersal of all these creatures sufficient to reestablish populations and the cycle can go on to be repeated in the next dry period.

This function is especially important in cases of several mosquito-eating fishes. In areas of the Everglades where Alligators have been completely shot out and where there are no canals or other fresh water, several species of mosquito-eating fishes have suffered for want of this natural dry season oasis that is provided by the valuable Alligator.

It is uncommon today for the Alligator hole to have such significant importance in the distribution and maintenance of wildlife, because of all the real estate lakes, golf course hazards and canals that have been excavated in this region. All such bodies of fresh water provide satisfactory habitat for the Alligator and many other species. However, a very fine natural Alligator hole may be seen on the Sanibel-Captiva Conservation Foundation Wetland Sanctuary. This is a very old hole that was once inhabited by a large Alligator since shot by poachers. I have introduced homeless Alligators to this hole eight times, but none has stayed.

This is a subject about which Richard Workman, Director of the Sanibel-Captiva Conservation Foundation, has written: "The ecology of an Alligator hole and the ecological role of the Alligator are yet to be fully understood or properly appreciated. It is noted, as an example, that the Alligator plays a key role in the distribution of nutritional elements essential to other life forms throughout the wetlands environment. Studies of crocodilians elsewhere have shown diminished fish populations after the reptiles were removed from the system."

In our region probably the main contributing factor to the declining fishery where Alligators have been completely eliminated is the lack of control on some predatory fish species which form a part of the Alligator's normal diet, thus allowing a disjointed overpopulation of such "trash" fish to the detriment of Bass and other desirable forms.

The greatest enemy of the Alligator is, of course, man. The early Florida naturalist, William Bartram, wrote some two hundred years ago: "The Alligators were in such incredible numbers, and so close together from shore to shore that it would have been easy to walk across on their heads, had the animals been harmless."

Since that day, man's decimation of the Alligator population has been unbelievably heavy. During the last century and the first three decades of this one, more than nine million Alligators were destroyed for leather and for meat throughout the several states of this species' range. In Florida, in one decade (1930-1940) more than one million Alligators were destroyed. By the early 1960's it became apparent that protection was necessary or this species would join the swelling ranks of those now extinct. Early protective laws caused the illegal leather trade either to disappear or go underground, but there were plenty of poachers, because good belly skins still brought eight or ten dollars per foot. Poachers of the sixties were virtually immune to the law, for local judges, themselves often products of a gator-hunting culture, would not hand down stiff sentences. Occasional fines could easily be offset against the high prices that the hides were bringing.

In the early years of this decade, the Alligator achieved federal "Endangered Species" status and it became a federal crime to transport them across state lines or to export the hides.

The Department of Interior states that "the amazing comeback of the Alligators shows what concerned citizens can do when they decide to help wildlife. The federal government was only the transmission mechanism for this successful recovery of an endangered species. It was the citizens and conservation officials of some southeastern states, insisting that local legislation be passed to protect the Alligator, which accomplished this."

In my opinion and in the opinion of other conservationists, the "amazing comeback" is something less than that. Alligator populations are nowhere near normal, but the fiction of their superabundance can probably

Why feeding of alligators is illegal.

Sanibel has a viable population of wild alligators, and most of them are harmless. For visitors to the island, these animals represent our most popular attraction. Of the 26 kinds of crocodilians that remain alive in the world today, the American Alligator is the most gentle and tractable. Yet this same alligator can be made into a dangerous animal by man's thoughtless artificial feeding and "taming." An alligator that is accustomed to being fed will become a potentially dangerous animal if left unfed for a period of time. Animals that are not artificially fed are less likely to become problems. A "tame" alligator becomes bold — he loses his fear of man. Some alligators have even been fed BY HAND. An alligator, given his intellectual limitations, has difficulty distinguishing between a piece of fish and the hand that holds it.

MECKLER B.

The feeding of alligators can be dangerous.
The feeding of alligators is a disservice to the animal.
The feeding of alligators is illegal.

Help preserve the most interesting and valuable of our wildlife species.

The alligator:

- attracts visitors
- patrols the rookeries
- provides water to all other animals in times of drought
- supports a healthy fishery

Southwest Florida Regional Alligator Association
P.O. BOX 241, SANIBEL

Sanibel poster attempts to promote detente between its primate and reptile residents.

be explained in two ways. After more than a decade without any legal harvest, the animal loses its fear of man and begins to offer a visible target, and game officials everywhere are constantly under stress from hunters seeking more and more target animals. This is one reason, I believe, that state game authorities and the Department of the Interior were willing to accept the fiction of an "amazing comeback" and were willing to "de-list" the Alligator from "Endangered" to "Threatened" status, thus allowing a certain amount of controlled killing. We are hopeful, however, that Florida will have stronger conservation regulations than other states, for Dr. Earl Frye, the chief game authority in Florida, has a more enlightened attitude toward the Alligator than can be found in most other states where the species ranges.

The other reason for the apparent "amazing comeback" is the fact that Alligators are more visible now than they used to be. With the destruction of natural habitat, the filling of sloughs, the bulldozing of forested areas, and other environmental insults such as the draining of vast areas to accommodate housing developments, the Alligator has had no place to go except into unnatural visible habitat, such as dug real estate lakes, ponds on golf courses, mosquito canals, drainage canals and farm ponds. Such artificial habitat renders the animal conspicuous because natural cover has often been largely destroyed. These artificial bodies of water make satisfactory homes for existing Alligators, since usually there are fish, turtles and other edibles, and Alligators get along very successfully in these locations so long as they are not molested. But such locations offer little nesting habitat. For nesting, the Alligator needs cover and seclusion. It's not common for nests to be found in unnatural exposed situations although such cases have been recorded.

The Interior Department based its de-listing from "Endangered" to "Threatened" status on an estimated 734,400 Alligators in all of the United States, of which they say there are 407,585 in Florida, the rest being in Louisiana, South Carolina, Georgia, Texas, Alabama, Arkansas and even Oklahoma (where there are officially *ten* individuals). These figures are not based on scientific censuses and are, at best, estimates. Our own estimate for Sanibel Island is 275 and I am the first to admit that this figure may be far from accurate. However the estimate is based on many, many observations.

The Endangered Species Program of the U.S. Department of the Interior is really beautiful in concept but in practice it is fraught with problems and questions. Certainly it was protective legislation that brought the Alligator back from near-extinction, but why is it necessary to fictionalize the degree of comeback and why is it necessary, once some success has been achieved, to place this species back into a classification which may permit it to be exploited? If indeed there has been some success in the recovery of this species, would it not be wiser to maintain its protected status, rather than to place it in jeopardy once more?

If Alligator hides find their way back onto the American and international markets, their sale can only stimulate the increased demand of crocodilian hides of all kinds and the other twenty-five or so kinds of crocodilians are in much greater jeopardy than is our Alligator.

In 1974, "Bismarck,"[1] a great Alligator of classical seventeen-foot length, according to Richard Beebe the ornithologist, was intentionally killed on Sanibel Island.[2] Such a shocking abuse of wildlife on a sanctuary island like Sanibel seemed to me intolerable. Seventeen different conservation organizations agreed, and we banded together to form the Southwest Florida Regional Alligator Association, which has been chartered by the State of Florida to undertake a number of activities. We educate the public to the value of this species in the environment. This activity, conducted by the Lee County Environmental Education Department under the direction of William Hammond, has produced a number of educational programs including one in Spanish entitled "Armando El Aligator, Esta Es Tu Vida." The educational program has met with considerable success and many youngsters of this region are learning to value this species.

---

[1]Many of our known individualized Alligators are named.

[2]On April 23, 1977, Dr. Frank C. Craighead, Sr., wrote me the following note about Bismarck: "It seems impossible that the 17-foot Alligator was destroyed for eating a dog. I saw that reptile, the largest I have seen in my 27 years in Florida. I couldn't believe my eyes."

Another undertaking of the Association is research—usually consisting of field studies, conducted on Sanibel Island by George Weymouth and elsewhere by Kenneth Shain of Edison Community College. We have a tagging program which enables us to monitor growth rates and migration patterns and make population estimates. Some startling records have been gathered of which I will mention one or two.

"Twiggy" was first captured by Charles LeBuff on August 11, 1972, at which time he was thirty and one-quarter inches long. Twenty-eight and a half months later he was captured again and was found to have grown an astounding five and a half feet during that period. This exceptional growth can be explained by his excellent source of food at The Island Beach Club where he had taken up residence and been fed large quantities of fish and chicken. Twiggy was thought to be female by people at the Island Beach Club, but later was ascertained to be a bull. Because of his being hand-fed, Twiggy had become a bit too bold and the management asked to have him removed. So we caught him and took him out to a far-distant point in the Darling Sanctuary, approximately seven miles from his home. A month later Twiggy was half-way back and had stopped traffic on the Sanibel-Captiva Road near the Elementary School. My son, David, and I captured him again and took him to the same distant place in the Darling Refuge. Three months later, Twiggy was back at square one, at the Island Beach Club where we trapped him again and took him to the same place in the Refuge. Two months and six days later he was back at his home pond. On July 4, 1975, Twiggy was captured and taken to Franklin Locks, up the Caloosahatchee River toward Lake Okeechobee, and released. He hasn't been seen since, but it would not surprise me a bit to have him show up again at his old haunts on Sanibel.

Twiggy displayed some unusual behavior. It's well known among people who keep captive crocodilians that freshly caught ones seldom eat in captivity for several months or until they become adapted to their new life style. On one of the occasions when we trapped Twiggy and were taking him on his ride to exile, he was in a large seven-foot-long wire cage trap. He had not consumed the entire chicken bait that had been used to recapture him, and the carcass was still hanging from the top of the trap. As my son and I drove with this trap strapped on the hood of his old car, Twiggy thrashed about quite a bit but in a moment of calm he spotted the chicken swinging from its wire and — amazing behavior indeed! — he proceeded to gulp it down.

Investigations have revealed many interesting migration patterns of Alligators. Also established is the fact that they occur in many different habitats. Although it is essentially a fresh water animal, the Alligator does go into both Gulf and Bay and can be seen with a certain degree of frequency in both habitats. In fact, we have ascertained that the Alligator can be found in all of the several habitats of Sanibel Island, even the high ground of the Mid-Island Ridge. I have, however, observed that Alligators held in salt water, without an opportunity to visit fresh water, will die, so always they must return to fairly fresh water sooner or later after their forays into marine situations.

The third function of the Alligator Association relates to problem Alligators. We are responsive to citizens' complaints and consider each case individually. If, in our opinion, the problem is imaginary or minimal, we endeavor, through discussion, persuasion and education to resolve the situation by persuading the complainant to withdraw the request. If we are convinced that a real problem exists and that the animal must be removed, we capture it and transport it to a location identified through the field study program to be a safe habitat. Before release, the animals are probably branded or tagged and full descriptive data are recorded.

We are kept quite busy with "problems" each spring and early summer for two reasons. It is the breeding season and males are visiting the various territories of females, seeking mates. One male may travel a considerable distance and may service several females.

The other reason for the abundance of Alligators at this time of year is the receding of fresh waters. For example, much of the water on the left or "fresh water" side of the "Ding" Darling Memorial Drive is saltier than the sea. This is a function of drying, as the mineral salt does not evaporate and thus concentrates in a reduced quantity of water. The Alligator seeks fresh water when there is an increase in the salinity of its existing habitat. Consequently, don't be surprised if you see an Alligator act-

Photos, Dallas Kinney,
Island Reporter

In addition to a nictitating membrane or "third eyelid" which protects the eye while permitting underwater vision, the Alligator has an eliptical pupil with the capability of remaining in a position vertical to the horizon, regardless of the animal's head position. This vertical orientation of the pupil provides uniformity of image.

ing like a lizard, running from the water side to the land side of the Sanibel-Captiva Road, as I saw one day in May, 1976. A lot of Alligators move around, but as soon as the rains come they again disperse and fade into a more appropriate and lower profile in our lives.

As you will recall, the only natural permanent body of fresh water on Sanibel is an Alligator hole and all the rest of the lakes and other bodies of fresh water are artificial. Thus it must be appreciated that natural habitat for the Alligator on Sanibel is very scarce indeed. Much of the slough has been built upon, filled and developed. The Alligator has been driven into the real estate lakes, canals and so on. These make very good homes for the Alligator, except that these real estate lakes usually have one or more houses built on their banks. And there lies the conflict. When, for example, a male animal starts his annual search for females, or other animals move in from salt-contaminated waters, overcrowding the remaining fresh water habitat, there is a problem. People tend to object.

"Marshall," a famous old twelve footer who lives in a lake on the edge of the Bailey Tract, demonstrates another kind of conflict. Marshall has been fed by many people throughout his long life, and when other Alligators moved into his territory, he objected. I tagged an eight-foot rival male who had one leg freshly torn off by Marshall. Another six-foot Alligator had his skull crushed by Marshall.

As the habitat is diminished by development, the remaining animals crowd too much together among their own kind and also find themselves too closely associated with hostile humankind. Hence the conflicts.

But we must not lose sight of the fact that Sanibel is a sanctuary island as well as a tourist island and that Sanibel's Alligators are attractive to visitors. So here on this Island, anything short of full protection is unthinkable. We have developed several methods for protecting the Alligator's interests as well as those of the human resident. Of course the most important one is to stop feeding these animals and thus not create so-called "tame" Alligators.

Another method is to construct a judiciously placed low fence. Many motels have ponds or lakes associated with them and in nearly all, there are Alligators which are attractive to motel guests. Fifteen or twenty feet of low fence, properly placed, will protect both tourists and Alligators.

Any such fence should be equipped with a sign stating that the feeding of Alligators is prohibited and why. If the area is too large to be fenced and there is a "tame" nuisance Alligator that has been fed, it is easy enough to discourage and "reeducate" him by throwing pebbles or poking a broom in the animal's face. But, of course, to be successful, feeding must be stopped.

It's not easy to tag an Alligator so that you are positive he will be marked for his entire life. Plastic or Monel metal tags are sometimes ripped off and a really satisfactory permanent tagging material has never been identified. So what we do is to mark or brand the animal by cutting tail crests. The Alligator has two rows of tail crests on the base or root part of his tail, and these become a single row distally at about the halfway point. These can be cut off, using a saw or a knife, making a permanent mark. For example, one could cut the second right tail crest and the twelfth left and have a code marking (R2, L12) that marks the animal for life. The main problem is that it is not easy to read such a mark in the field, even with binoculars. Consequently in order to identify a branded Alligator, you often have to catch him and that is a lot of trouble.

A couple of years ago, Charles LeBuff and I were catching an eight-foot Alligator and we had the process pretty well under way. The animal was in ropes and I was standing on his head while Charles started to bind the jaws shut with tape. At this juncture, the beast threw me off and when I landed, my foot was in his mouth. Fortunately he did not roll or I might have lost my foot. Instead, Charles and I were able to induce him to release it and only moderate, though excruciatingly painful damage resulted. This Alligator, Charles and I agreed, was the toughest and most irascible animal either of us had ever seen. This trying experience led us to develop safer capture techniques.

Usually real problem Alligators are accessible. Either they have been fed and are tame, or they are wedged in under a parked car or in somebody's pool or patio or in some such accessible situation. If they stay at a distance out in a pond, have not been fed and are not easily approached, they are not, in our opinion, problem animals.

So in order to catch a problem Alligator, we have developed the following system: a steel cable noose is used as a snare, and is mounted on a pole so that you can reach eight to ten feet toward the animal and drop the noose around his neck. Then, with a steady pull on a strong nylon cord, you can haul the thrashing beast out of the water to a convenient open area such as a field or parking lot. There, two more heavy nylon nooses are substituted for the snare which is then recovered. These nooses are equipped with snaps so that later the lasso can be immedately released. The animal is next stretched between two trees or two cars or other solid anchors. Usually he will roll. Alligators and crocodilians in general defend themselves, tear up their prey and subdue their attackers by rolling over and over again. In this case the rolling habit defeats the Alligator's purpose for it tends to bind itself up in the ropes.

Using a long-distance method of lassoing the upper jaw and tying it to the lower jaw with fine cord, the mouth is rendered temporarily harmless after which heavy tape is applied to the jaws. This, as you probably know, subdues the creature because the power in the jaws of an Alligator is in the closing and not in the opening. Therefore a relatively fine nylon cord or masking tape can effectively hold the mouth closed.

The Alligator tail is a vastly overrated weapons system. We tend to ignore it entirely and none of us has ever been seriously injured by an Alligator swinging its tail.

The Alligator by this time is rather subdued and traumatized and we can easily transport him by car or truck to the place where he will be released. By that time he's really in a fatigued condition and can easily be tagged or branded as described above. His sex, size, measurements, etc., are recorded and the beast is freed.

Releasing an Alligator can be almost as dangerous as capturing one, so great care is taken to make sure to be absolutely clear of the head at the moment when the last bonds are removed. Only the Alligator's now-fatigued condition serves to render this operation a little bit safer. A straightened and blunted gaff hook is used to remove the last noose and the animal is urged into the water.

Reptiles are divided into several major groups: snakes, lizards, turtles, the Tuatara of New Zealand and the crocodilians. There are only about two dozen different kinds of crocodilians and our Alligator is perhaps the least stressed of all. It is no exaggeration to say that all crocodilians are in serious danger of extinction.

The crocodilians, the birds and the dinosaurs had common origins about two hundred thirty million years ago in a now-extinct group called the Thecodonts. As we all know, some decendants of this ancient lineage have long since died out, there being no dinosaurs left in the world today. However the crocodilians share with the birds this ancient Thecodontian origin and both are still much in evidence.

Of the crocodilians, there are two major groups—the true Crocodiles and the Alligators and Caimans and a minor third group consisting of the Gavial which we won't consider here. The Alligators have evolved in the last 36 million years from the Oligocene period until the present, largely on the American continent. There have been several evolutionary dead ends, but two Alligators remain in the world today — our own *Alligator mississippiensis* and the Chinese Alligator, *Alligator sinensis,* which in the Pliocene, some thirteen million years ago, crossed the Bering land bridge to China. The Chinese Alligator has been known since the dawn of history and is said to be the animal that led to the concept of the Chinese dragon. This exceptionally rare animal, which lives in the Yangtze River basin of China, is almost extinct. There is little current knowledge of any that may remain in the wild. There are, however, forty or fifty captive animals that are in the West and an effort has been made to establish a few in a breeding colony in Louisiana under the auspices of members of the Crocodile Group of the International Union for Conservation of Nature. It is hoped, but the prospects are not good, that this effort will result in at least a limited population so that this species will not disappear completely from the face of the earth.

The breeding, nesting and parental care habits of Crocodilians have recently been studied—first by scientists at Everglades National Park. The principal worker, John Ogden, discovered some fascinating maternal behavorial patterns of the Florida Crocodile, *Crocodylus acutus.* Later work with the Nile Crocodile, *Crocodylus niloticus,* showed similar patterns. This work with both species demonstrated that the mother Croc actually guards

We heard that an Alligator had fallen into an elevator shaft foundation. →

← *So we went to take a look.*

Sure enough — there he was. →

He was then hauled away to a safe habitat.

Photos, Peter Larson, *Island Reporter*

the nest and when it is time for hatching, signalled by croaking from inside the egg, she actually digs down and gently removes the eggs which she helps open with delicate maneuverings of the powerful but gentle mouth. She then gathers the young in her mouth and carries them to water where they are released. The juveniles stay together and are guarded by the parent for some time. More recently this behavorial pattern has been confirmed in the American Alligator, the Spectacled Caiman, *Caiman crocodilus,* the Saltwater Crocodile, *C. porosus,* and Australia's Johnson's Crocodile, *C. johnsoni.* This writer therefore feels confident that this behavior pattern occurs through all Crocodilian species. One is only amazed that its discovery is so recent.

Alligators hatch from their vegetable mound nests in August or September and as stated above, a similar pattern of nest and egg opening and later parental care has been confirmed. The newly hatched little ones are less than a foot long and grow very rapidly—about a foot a year. That is, the ones who don't fall prey to Herons, Otters, Turtles, raccoons and other predators.

At about eight years of age, the Alligator should be about eight or nine feet long. They achieve sexual maturity at six years and usually live to be about twenty-five years old. A fifty-year-old animal would be very old indeed.

On August 16, 1975, I studied a recently-hatched Alligator nest on Sanibel. The condition of the youngsters and their location, together with the dampness of the egg shells, led me to believe that the eggs had just hatched the day before. When I was a youngster living near here and studying the Alligator, I don't recall any hatch earlier than September. But on this occasion, hatching was two weeks earlier and the babies were a bit longer than those I used to see in my youth.

One could see where this great mound of vegetation and sand had been dug out by the mother when she heard the young croaking from inside the egg. Young Alligators are very vocal, even before hatching. The mound was perhaps two and one half feet high and five feet in diameter. There was a cavity in the top from which the baby Alligators had been removed and which was perhaps 14 inches deep and two feet in diameter.

Fortunately for me the mother was not to be seen, but before I caught the little fellow pictured here, you can be sure I looked around carefully because occasionally a mother Alligator will protest rather strenuously any interference with her young.

This nest was located on a spoil mound in the central fresh water wetlands of Sanibel. Nearby were many Red Mangroves and Buttonwood trees. This is rather unusual Alligator nesting habitat, for the Red Mangrove is usually a salt water species. However, many hiding places exist among the reticulated prop roots of the Red Mangrove, not only for baby Alligators, but also for numerous other animals. I was attracted to this location by the sound of the babies calling. When they heard me crunching through the dry sticks and leaves, they started their juvenile distress call which is sort of a soprano croak or grunt. When the first one spotted me and called out, this stimulated all the rest of the dozen to follow suit. This croaking call was very obvious for perhaps 40 to 50 feet. They were shy and dispersed when I moved, but when I sat on the bank imitating a stump, they soon lost their fear and some swam near my feet.

I enjoyed an hour of quiet observation as they moved about the area, always in voice contact with each other. If I moved or swatted a mosquito, of which there were many, they would take off from near my feet and head for deeper water. But as soon as I was still again they felt the shore was safer than the deep water. I'm inclined to believe they were right because of the many Snapping and Soft Shelled Turtles and Anhingas which will feed on baby Alligators. So when all was still, they would head for the shore where they hid again in the roots of the Mangroves.

I believe the communication of the alarm cry among such a group provides important protection. It is well known that only a few babies survive to adulthood, but the cry of a baby—even when being swallowed by a Heron—provides added safety to all the remaining members of the clutch. Thus I believe that the practice of capturing a nest of babies and moving them to what we humans believe to be a "better" habitat with greater food supply, may be a real disservice and may actually defeat the conservation effort.

The one I captured was marked and will be identified for life. We may pick it up again in a couple of years and thus have another record of growth and movement.

This little fellow was exactly ten inches long. Its umbilicus was hardly closed, but it was already a spunky little miniature monster. It was very ready to bite my finger and hold on to the extent that it could be lifted up in the air, supporting its weight with its powerful bite. Our ten inch monster was about an inch longer than most hatchlings that I have seen or found recorded in the literature. Our Sanibel Gators seem to be precocious!

This little specimen was returned to the Sanibel slough and hopefully is doing its thing, eating the little mosquito fishes, *Gambusia,* as well as insects and any other small animal that crawls, swims or walks its way.

Baby Alligators are much brighter in color than adults. The yellow and black markings are very clear and bright and readily apparent, but in time they will fade.

Since our specimen started off at almost a foot and will grow about a foot a year, in eight years it should be almost nine feet long. After that, the growth rate is somewhat slowed until the animal reaches about twelve feet, and after that, further growth is achieved very slowly. But there are records of these monsters reaching an unbelievable length of nineteen feet.

I keep referring to this specimen as "it" for upon hatching, the Alligator's sex, unbelievable as this may seem, is undetermined. Sex determination in Alligators is influenced

Although only a few hours out of the egg, this baby monster can support his weight with his already powerful jaws.

Photo, Mark Twombly, *Island Reporter*

by the environment and actually happens *after* it has hatched and is swimming about. Temperature and, I believe, disturbance play an important part in determining sex.[1]

For those of us who got our sex determination genetic concepts from the old "XX" and "XY" school of thought, it comes as a shock that something "new" like this could have been going on for one hundred eighty million years, ever since the Mesozoic—and we mammalian newcomers have only now made this discovery.

It has been established that if you disturb an Alligator nest, take the eggs, carefully hatch them, feed and rear them, you will get all — or almost all — one sex. If you leave that nest alone, let nature take its course, you will get normal sex distribution between males and females.

Some efforts have been made to artificially rear rare Crocodilians for release into the wild. This "Operation Head Start," as some have called it, was once, to my mind, the most successful captive breeding technique of all for Endangered Species. But now doubt is being cast upon its efficacy—perhaps such efforts are producing all one sex.

And there is another factor rearing its ugly head: I call it the "Elsa Factor." Remember how tough it was for Joy Adamson to educate tame lions, Elsa and her cubs, and return them to the wild? This problem was stressed in her "Born Free" series of books.

Many conservationists who interested themselves in captive breeding as a means of species preservation felt the "Elsa Factor" to be related only to higher animals—felines, Orangs, Chimps, etc. But now we are finding that "lower" animals such as the Alligator, when raised in captivity, have the same sorts of orientation problems that Joy Adamson observed in Elsa: they can't take care of themselves in the wild. Or at least most can't. A crococilian raised in captivity and released, usually dies.

A captive bred crocodile (or Orang, or Prezwalski Horse or Wisent) looks like the real thing, but isn't. Changes, often genetic, sometimes behavorial or physiological, (or as we now know with crocodilians, sexual) have taken place that threaten the animal's ability to survive and reproduce in its ancestral habitat. Breeders of Endangered Species are naturally terribly concerned about this and many are abandoning this "second" method of species preservation and reverting to the primary method—habitat preservation with protection of the animals in the wild. But with burgeoning human populations everywhere in the Third World, this method too seems bound to fail in the long run.

So much for Sanibel's most noteworthy wildlife species. It is my hope that even though protection is withdrawn from the Alligator in other parts of its range, we can keep the species completely protected on this Sanctuary Island. It is one of Sanibel's noblest traditions that wildlife is highly regarded here and rigidly protected. Perhaps this discussion of the Alligator can help to instill in the visitors and residents of Sanibel an understanding that can lead to even more stringent protection of this most valuable ancient species.

---

[1]In a related phenomenon, temperature determines sex in the Common Snapping Turtle (*Chelydra*). Eggs from a clutch incubated at 20°C or 30°C will form all females. Eggs from the same clutch incubated at 24°C will produce 100% males; at 26°C, 99% males. Temperature has been proved to influence sex determination in other vertebrates also — other turtles and a lizard. Dr. Chester Yntema of the Upstate Medical Center, Syracuse, New York, informed me of these findings after my correspondence with him on Alligator sex determination.

Marshall is one of our best known Alligators. At twelve feet, he is also one of the largest.

6 *THE ARMADILLO* A Success Story

The Nine-banded Armadillo is very well-known on Sanibel Island. Every summer many Armadillos are seen, mostly smeared on the road by traffic. A few may be seen snuffling around the road edges looking for prey —usually at night, for in summertime, this is a nocturnal beast.

The Armadillo has not always been such a prominent member of Sanibel's fauna. Dr. James N. Layne of the American Museum of Natural History, who has provided me with many mammal notes for Sanibel, indicates that the Armadillo appeared here only after the opening of the Causeway in 1962. The first time it was mentioned by the "Ding" Darling Refuge personnel was in 1964 when they stated that several competent Armadillo observations were made on Sanibel Island.

Ernest P. Walker, a prominent mammalogist, stated that it is a swimming animal. Others claim that it walks along the bottom of streams or canals. So there is much controversy. But it is probable that both are correct.

In August 1974, part of this question was settled to my satisfaction when Alan Pote found an Armadillo swimming near the channel markers toward the middle of Pine Island Sound between Captiva and Pine Island. Molly Eckler Brown happened to be present at Timmy's Nook dock when this animal was brought ashore. It was exhausted and completely worn out, but as far as anyone knows, it did survive the experience after it was released on Captiva.

Swimming comes rather easily to this heavy-looking creature because it has the faculty of swallowing air and thus inflating its digestive tract, including the intestine, rendering it buoyant enough to float.

The other theory—that the Armadillo walks on the bottom of water bodies—has been demonstrated to be at least partially true, In the laboratory, it can be shown that an Armadillo will sink to the bottom of a tank and proceed to walk along, so it's possible that for long stretches, swimming is accomplished by means of the inflated G.I. tract, and for short canal crossings, the animal may simply walk across the bottom.

In light of this evidence, the explanation of this species' appearance on Sanibel at the time of the opening of the Causeway in the early 60's is, I believe, purely coincidental. Charles LeBuff, who has worked with the wildlife on Sanibel Island for a very long time, was told that Armadillos were initially released on Captiva at about that time. Records also indicate that some were released on the east coast of Florida in the 1930's and gradually expanded through much of eastern, northern and central Florida in the succeeding years until the 1950's when it was found in most regions of peninsular Florida, except for the Everglades region and southwest Florida.

At the same time, the Armadillo was expanding its natural range from Mexico. It moved north in the middle of the last century to Oklahoma and Kansas and east to the Gulf states over to and including the Florida Panhandle. Therefore, about the same time the introduced population was spreading throughout the peninsula, so was the naturally-expanding population extending downward from the Panhandle to southwest Florida. Given the animal's proven ability to swim, plus the rumored introduction of specimens to Sanibel and Captiva, it is not surprising to me that we do, indeed, have a thriving Armadillo population today.

The Armadillo belongs to an order known as Edentata, a name meaning "without teeth." The ways of scientists are sometimes inexplicable, because not only are most edentates amply provided with primitive peg-like teeth, without enamel, but one species, the Giant Armadillo, *Priodontes giganteus,* of South America, actually has more teeth than any mammal other than some whales.

Scientific nomenclature is full of such contradictions. This is explained by what is called the law of precedence, which simply means that if a taxonomic worker of yesteryear applied a name to an animal or group of animals which is later proved to be factually nonsensical or not really descriptive, the name will persist because it was first. Another example of this is the Florida Black Racer, a well-known snake on Sanibel. This snake is called technically "*constrictor*". In fact it does not constrict, yet the name persists in spite of the error of some early zoologist.

New World Edentates include the sloths and the anteaters as well as the armadillos. These animals, except for the subject species, are found exclusively in tropical areas. Our own Nine-banded Armadillo does range into colder climates, such as Oklahoma and Kansas, but it suffers greatly during extremely cold weather, and large numbers have been known to die. Lacking hair (or rather fur, for there are a few bristly hairs on armadillos), the animal is highly susceptible to cold and must regulate its body temperature by means of behavior rather than insulation. Thus in summertime, Sanibel's Armadillos are largely nocturnal, for the night temperature is quite warm. In the winter, when even South Florida experiences chilly weather, it is not uncommon for the Armadillos to seek their prey in the daylight hours.

The Nine-banded Armadillo is *Dasypus novemcinctus mexicanus.* "*Dasypus*" simply means hairy-footed; "*novemcinctus*" denotes that it is wearing nine "belts"; and "*mexicanus*" identifies the sub-species first described from Mexico.

This same nine-banded species ranges well into South America, where some races or subspecies are considerably larger than our own. The National Museum in Rio de Janeiro has a specimen of approximately twice the size of our Armadillo.

The word "armadillo" is a word of Spanish origin and refers to the armor-like outside covering which the animal possesses. The dorsal bony carapace is composed of three parts and forms a complete curved shield around the body. The foreparts are known as the scapular shield. This is the shield which covers the shoulders. The pelvic shield is the hindmost. Between the two shields are the nine narrow transverse bands or belts, each with sixty or seventy closely-knit hard elements known as scutes. Along all this armor are a few flimsy, spindly hairs growing from the soft connective tissue which attaches the bands one to another and which permits a sort of accordian-like extension and compression of the back. Actually this hard armor is really skin, remarkably modified to provide this protection, but skin nonetheless. It is not in any way connected to the skeletal system.

The Armadillo is not particularly noted for its intelligence. Its responses are largely instinctual, as one would assume in an animal with such an anatomically primitive brain.

In recent years the Armadillo has been used as an experimental model by the biomedical research community in the fields of experimental surgery, reproductive biology, skin transplantation, caesarian section and as a tool in the study of immunization.

About fifteen years ago I learned that the Nine-banded Armadillo was being used as a biomedical research model in Carville, Louisiana, the site of the only leprosarium in the continental United States. So you can imagine my surprise in reading, just a few months ago, that the bacterium which causes leprosy, or Hansen's disease, *Mycobacterium leprae,* had been discovered in wild populations of the Armadillo. One could speculate that somehow infected animals escaped from Carville and joined their wild brethren who were increasing their population over an ever-greater range, and that this bacterium thus became associated with the species. This is pure conjecture, of course, and probably a more logical theory relates to the normal low body temperature of this almost hairless animal which provides an excellent environment for this bacterium. Leprosy, as everyone knows, attacks human extremities—the parts of the body which are not sustained at the normal body temperature. I am sure that the appearance of this bacterium in the Armadillo species will be the subject of intensive study which will contribute a great deal to our knowledge of this fascinating creature.

Another interesting aspect of the biology of the Armadillo is that it almost always gives birth to four young at a time, all derived from a

single egg cell, and each, therefore, having an identical genetic makeup. Polyembryony, as seen in the Nine-banded Armadillo, is rare in mammals. Among Armadillos, there are exceptions to the rule of four, but they are rare.

In southwest Florida, breeding occurs in July and August. The fertilized egg cell of the Armadillo may remain in a resting stage for three or four months, or even longer, before it becomes attached to the uterus, after which another three or four months of development may occur. Thus the period from mating to birth can be as long as nine months. However, unlike that in the more complicated animals, such as the primates, fetal development is not occurring throughout this entire lengthy period; the delayed implantation in the uterus tends to make the three or four-month true gestation period appear much longer.

There are quite a number of other armadillo species in South America, the most noteworthy being the Giant Armadillo, *Priodontes giganteus,* already mentioned. This creature may grow to four feet and be the size of a hog. It is very rare today and is on the International Union for the Conservation of Nature official endangered species list. In my youth in South America, I remember seeing this species in British Guiana savannah country although even then, forty years ago, it was scarce.

Someone once gave me a "shell" of this species. The three-foot long shell weighs thirty pounds, a fact I can attest to, having moved it along with myself and family several times over the years. The Giant Armadillo has met almost total extinction at the hands of man, virtually its only enemy. Unfortunately, most rural South Americans do not understand wildlife preservation. A Brazilian *caboclo* will almost always behead any animal first and ask questions later. The opening up of Amazonia through a network of roads will spell the doom of this and many other species.

The smallest armadillo is the Fairy Armadillo which may be only six or seven inches long. This lovely little animal bears the rather cumbersome name of *Chlamyphorus truncatus* and lives in Argentina. It is largely a burrowing creature and is equipped with much more hair than its cousins.

The Armadillo is not popular with many Floridians because of its tendency to tear up grass, interfere with agriculture and disturb plantings, in its never-ending search for insect prey. It also consumes a few good reptiles which, from my point of view as a herpetologist, is not desirable. On balance, however, I feel it to be a useful animal and in spite of speeding automobiles which take their toll daily on our highways, the Armadillo's polyembryonic fecundity will, I believe, compensate for the road kills and enable this most interesting creature to survive in spite of us.

7 *SANIBEL'S MARSH RABBIT*
Don't Look for a Cottontail

Sanibel has a charming little animal in the Marsh Rabbit. Its technical name is *Sylvilagus palustris paludicola.* Our rabbit is most easily distinguished from other rabbits in Florida by the fact that it does not display the familiar "cotton" white tail. Its upper parts are blackish or reddish brown. The underside of the tail is brownish—definitely not white. The ears are short and so are the hind feet.

Marsh Rabbits are aquatic in habit. They are known to enter water, either salt or fresh, and swim skillfully for distances of a thousand feet or more. As brackish swamps and the edges of mangrove forests are inhabited by this species, an efficient swimming ability is obviously an excellent protective adaptation against storm tides.

If you want to see Marsh Rabbits, look in home gardens, along roadsides and the "Ding" Darling Wildlife Drive as you search for your favorite birds. They are usually seen in ones or twos. Their "signs" are everywhere visible and as far as I am able to ascertain, the Marsh Rabbit is successful on Sanibel and is in no immediate danger, although there are several pressures.

One of the problems the Marsh Rabbit faces here is the feral domestic cat. For some reason, people in Southwest Florida think of Sanibel when they want to get rid of a litter of kittens, when they really should be calling the Humane Society. These animals consume many Marsh Rabbits.

Another problem is the ever-increasing automobile traffic. But the Marsh Rabbit handles this rather more successfully than do Raccoons, Opossums and Armadillos. Hundreds of those animals are killed every year on our roads, but I have seen very few Marsh Rabbits lying dead on the roadways.

A lot of people think of rabbits as rodents and so thought mammalogists for many years. The group that contains the rabbits, hares, and the pikas, is called Lagomorpha. For many years this was considered to be a suborder in the great order of Rodentia. More recently, mammalogists have included the lagomorphs in a separate order and they are no longer considered to be rodents.

Moreover the fossil record shows the lagomorphs to have separated from the rodents as long ago as 60 million years back. Also, modern blood tests have been employed to endorse or refute the more commonly employed methods of arriving at phylogenetic relationships and they have shown there to be no close relationship between the lagomorphs and the rodents. It has thus come to be accepted that if they did have common ancestry, it was a very long time ago.

There are only two families in this big order and they are the Ochotonidae, containing the pikas, and the family Leporidae, containing all the hares and rabbits.

The terms hare and rabbit are a little like the terms turtle and tortoise, or dolphin and porpoise. They don't have much difference in meaning. Rabbits and hares all belong to the same family and many mammalogists today tend to discount any differentiation between those two terms. As a general rule, vulgar names in zoology have little meaning anyway, so I'm not going to waste any time trying to decide which are hares and which are rabbits.

Rabbits show some similarities to the hoofed mammals. Some of them sort of "chew their cud." Only instead of regurgitating as a cow does, the rabbit simply eats his own feces. The pellets are of two kinds, moist ones and dry, hard ones. It's the moist ones which are "recycled" as it were. So in the case of a lot of rabbits, most of their food goes through at least twice. Certainly they are right in the current swing of things in this age where

recycling has become such a popular subject.

Rabbits can be found nearly everywhere in many habitats on most continents of the world. They were absent from Australia and New Zealand until a nostalgic Briton introduced them there, much to the regret of his grandchildren. It can be safely generalized that the introduction of any exotic animal or plant into a new environment is a potentially dangerous thing. Thus it was when Thomas Austin felt the need to have English rabbits on his Australian farm. He should have had second thoughts. In 1859 this gentleman brought in a dozen pairs of European Rabbits to his farm in south Australia. These few animals developed into a huge and destructive population that was to cost billions of Australian dollars and fantastic amounts in terms of environmental destruction.

New Zealanders soon followed suit. They failed initially but eventually they also introduced the European Rabbit, with similarly devastating results in their country.

Rabbits eat about a quarter of their weight in vegetation every day. Thus in both Australia and New Zealand it was ascertained that half a dozen rabbits would consume as much as one sheep. An area that had originally sustained 50,000 sheep would only hold, two years after the rabbits moved in, 5,000 sick and undernourished sheep. In this manner, Australia and New Zealand were forced out of the sheep business and into the rabbit business. Billions of rabbits were frozen and shipped out of those two countries over the years. Rabbit pelts were exported by the thousands of tons. Felt hats consumed much rabbit fur, but felt hats don't bring in the foreign exchange that wool can earn, and Australia and New Zealand suffered huge losses.

In one year in one Australian state alone, more than 20 million rabbits were eliminated, but still they came. Fences were put up in an effort to contain rabbit populations, but this didn't work very well, for the European rabbit, unlike our own, is a great burrower.

For the better part of a century, Australia and New Zealand were dominated by rabbit problems. Then the Australians got a break from South America, where it was learned that the indigenous population of rabbits were carriers of a disease to which they were immune, but a disease to which the European Rabbits were not immune. This is called *Myxomatosis cuniculi.*

The virus was imported into Australia and introduced to the rabbits of that continent after it was ascertained that neither the native animals of Australia were affected by *Myxomatosis cuniculi,* nor were the Australians themselves. This diabolical plot worked very well indeed, as hundreds of millions of rabbits began to die all over Australia. The reduced rabbit populations in Australia and New Zealand are now of a magnitude that people can cope with and still maintain their agricultural economy. But how much better it would have been had the original introduction not taken place.

There is a lesson here for us in Florida, because Florida is potentially a disaster area from exotic introductions. There have been more than two hundred species already introduced to the state and some of them could grow out of all proportion and achieve a magnitude that could be devastating to our agricultural economy.

Fortunately our little Marsh Rabbit belongs here and is a desirable member of our faunal community. So enjoy him and don't be too concerned if once in a while he nips one of your choice plants in the bud. After all, Sanibel is a wildlife sanctuary.

8 *A DAY IN THE LIFE OF THE "DING" DARLING REFUGE*

The J. N. "Ding" Darling National Wildlife Refuge includes four tracts on Sanibel but here we shall discuss only the main Refuge tract consisting of bayous and mangrove forest that is transected by the "Ding" Darling Memorial Wildlife Drive.

Each year thousands of visitors join with local residents in driving this five-mile one-way route, most of which is a dike road through mangrove forests. Those who take the time to study and to learn, to look and to appreciate, will find here a treasure house of nature.

As with all of nature's scene, it is ever-changing. I will attempt to describe here only what caught my eye on one late winter morning during a leisurely trip through the Refuge. On hundreds of other occasions I have seen other sights and heard other sounds, and so will you if you look and listen.

Very early in our trip we enter the mangrove forest. On the left, on the far side of the borrow ditch, can be seen some White Mangroves. On the right side are principally Red Mangroves — many youngsters, but also older trees.

Walking down the left, or so-called "fresh-water" side of the dike, is an all white Little Blue Heron. This seems to be a contradiction in terms, but the bird is a juvenile. As it gets older, it will attain the slate-blue color of the adult, and live up to its name. Both adult and juvenile have two-toned bills—the tip appears to have been dipped in an inkwell.

It's sometimes hard for the novice to distinguish between the large white birds, so you must know the characteristics of the various species. The Little Blue, as an adult, is almost all blue. The Louisiana or Tri-Colored Heron is slightly larger than the Little Blue, and it has a white belly. A Snowy Egret, with his yellow feet, his black bill and his completely white body is easily distinguished from the Great Egret, formerly known as the American Egret, which is black-legged and yellow-billed and somewhat larger, but also with a white body. Cattle Egrets, White Ibis, Wood Storks, white phase Reddish Egrets and, of course, the Great White Heron, which is seldom seen here these days, add to the confusion.

Swimming past on the left side is a Double Crested Cormorant with his hooked bill. His very skillful diving ability vies with that of the Anhinga, or Water Turkey, which can also be seen diving for the fish that sustain him. Cormorants chase their prey. Anhingas, on the other hand, set underwater ambushes much of the time. The Anhinga's bill is sharp-pointed and useful as a fish spear, but it also sometimes grasps the prey.

The Brown Pelican fits his popular name only when juvenile; as the adults achieve mature plumage, one wonders why it is reasonable to consider them "brown" pelicans since the elaborate adult coloration—the white head, brown neck and light wings— is highly noticeable.

The Pelicans in these shallow waters do not dive, as they do out in the deeper Gulf. The water is so shallow it would be dangerous. Consequently, shallow dives and float fishing for smaller fishes is more the order of the day.

The Red-breasted Merganser exists on the "Ding" Darling as an almost entirely female population. Most of the males take separate vacations farther to the north.

The Mergansers demonstrate almost daily their symbiotic, or cooperative, relationship with other species of birds. The Mergansers will travel the borrow ditch and drive the *Gambusia,* and other small larvivorous fish they don't succeed in catching to one side, while the herons, such as the Louisiana Heron and the Snowy Egret, will march along the side of the ditch spearing a few, catching a few and driving the others back toward the Mergansers. Thus, perhaps, the origin of the expression "poor fish" — there is no safe place for the poor fish to turn; he flees from the Merganser into the mouths of the Herons and back again, losing fellow-members of his population at each encounter.

It's about 7 A.M. and we have stopped by station number two. The dawn has just broken. To the east, a spectacular sunrise can be seen across the Red Mangrove forest. To the west, the new light of day is shining on a few birds that can only barely be seen. They do seem to be pink, but at this time the light can play tricks on one's vision. Yes, they are indeed pink—they are Roseate Spoonbills standing on the flats just beyond the borrow ditch. This is rather unusual, for in other years Roseate Spoonbills did not spend the winter in Sanibel. To Spoonbills, Sanibel is way up north. Most of their fellows have gone farther south to tropical America to breed. One of our great local ornithologists, Dick Beebe, believes that these five are actually youngsters spending their first independent winter here instead of following the flock to the southern breeding grounds.

Soon we will begin to see a greater number of Roseate Spoonbills as they return to spend the summer with us, and we can expect to have 400 or more. Our flock is a very large proportion of the U.S. Roseate Spoonbill population—maybe 15 to 20 percent, or even more.

As the light gets brighter, we begin to see a number of ducks on the water to the left of us. There are Lesser Scaup, Wigeons or Baldpates, Pintails, Shovelers and the one we have with us all year round, the Mottled Duck or the Florida Duck. The latter is called the "Texas Duck" in Texas, so you can call it what you will. Both sexes look somewhat like the familiar female Mallard.

There are several Terns visible in this early morning light and we see a few Black Skimmers cruising along with their delicate shallow

Louisiana Heron

wing beat very near the surface of the water, knifing the surface with their long lower bills and picking up by feel the *Gambusia* and other small fish.

On several occasions I have referred to the "borrow ditch," a term widely used in these parts. It simply means that the soil was "borrowed" from the ditch to build the dike. For this reason, on both sides of the dike road we find rather deep water, but beyond the ditches the water is very shallow and on a low tide, many acres on the right side are actually exposed.

The left-hand or "fresh-water" side is really not all that fresh. It was filled in 1975 at the end of the dry season by a high tide and the tidal opening was then plugged. Consequently there remains a great deal of salt water impounded inside. This, of course, is diluted with the rains of each summer and fall.

Parts of this water area today are somewhat less saline than the Bay or tidal water. However, between the late winter months and the time the rains start again, evaporation occurs and unless the impoundment is allowed to flush by opening up a tidal outlet under the road, it will gradually become more and more saline.

High salinity in 1975 and 1976 is believed to be the reason for Sanibel's small duck population. Much of the fresh water vegetation that the "puddle" ducks consume was prevented from growing by the too abundant salt in the water. This might be corrected by means of opening the pipe on a very low tide, sealing it again after flushing, and letting the rains accumulate, eventually providing somewhat fresher water than we find today. Also there is a long term but unfortunately underfunded Federal plan to provide water control structures through the dike to provide better salinity control in one of the impounded cells.

Now let's go on to station number three and observe the vegetation. There are four types of mangrove trees in the mangrove forest: the Red Mangrove, the White, the Black and the Buttonwood. These are not all true mangroves. They are designated "mangroves" because of their ecological inter-relationship — they all live and work together at the edge of the bays and are often inextricably involved, one with the other, in this complex system.

The Red Mangrove, the only "true" mangrove in Florida, is considered more adventuresome than the others. This is the one you see so abundantly on the right hand side of the Dike Road, having produced the islands through the collection of vegetation, mud and debris among its prop roots. Among the Reds are a few tall Black Mangroves.

Aside from land-building, the Red Mangrove is a main source of nutrition for this whole ecosystem. The brown dead leaves, technically called mangrove detritus, fall into the tidal water and begin their decomposition, with the aid of fungi, protozoa and bacteria. As this process develops, the protein values of the mangrove leaves increase. Other small microscopic species that consume the leaf and also consume the consumers of the leaf are, in turn, eaten by larger creatures and so it goes on up the food web to the summit, where the predators reign supreme—fish, reptiles, mammals and birds. The Great Osprey and the Bald Eagle ultimately, to a large degree, depend on the Red Mangrove leaf for their nutrition as do the Snook, Redfish and other valued fishes of our waters—not to mention the Alligator and our all-important Pink Shrimp.

Eric Heald, a noted botanist, has documented the values of the Red Mangrove so that all can understand its tremendous importance. Environmentalists and ecologists everywhere admire the contributions of this most valuable of all our trees. The Red Mangrove is almost sacred to Florida nature lovers and hence I have resolved to change its name to suit my fancy. Instead of *Rhizophora mangle,* as it is properly called, I have renamed it *Rhizophora religiosa,* the Sacred Bo Tree of the environmentalist. Under it, naturalists can sit and ignore their wet pants while they contemplate the intricate wonders of nature.

Flowers, seeds and seedlings can be seen on the Red mangrove simultaneously. The seedling actually germinates from the fruit or seed while it's still on the tree, a process known as viviparity. So you see, people and other mammals are not the only viviparous living things in the world.

There are three other "mangroves," including the Black, which you can always tell by the salt that accumulates on the leaves. On a dewy morning you can see dampness and on a dry, sunny day you can see the crystals of salt glistening in the sunlight. This salt is harvested in some parts of the world. In Nigeria and India, it is used as table salt. Black

Roseate Spoonbills

Mangroves are also equipped with gaseous exchange organs—pneumatophores—the pencil-like structures that rise above the mud around them all. They also provide firmness to the whole tree which often lives in a very unstable location. Del Hicks, an Environmental Protection Agency Scientist, has studied the nutritional values of the Black Mangrove and finds this species to be of as great importance as the Red Mangrove.

The White Mangrove has two salt elimination glands on each of its petioles or leaf stems—just tiny little bumps that serve to identify this common tree which has light-colored bark and a rounded leaf, often with a heart-like indentation on its tip.

Great Egret

The Buttonwood Tree is the least adventuresome of all the four mangroves and occurs on higher lands. It has a pointed leaf and is also equipped with two petiole salt glands. Its rough bark serves as fine habitat for orchids and bromeliads, which live epiphytically on many of these trees.

Often on the side of the road, you may see little Ground Doves seeking grass seeds. I prefer the more picturesque West Indian name, "Tobacco Dove." This is our smallest and most abundant dove today. In flight, these pretty little birds display a reddish flash on the wing.

In past times, the White-crowned Pigeon was very common on Sanibel. Today the White-winged Dove occurs on Sanibel occasionally. The Mourning Dove, although abundant on the mainland, comes here rarely.

Early in the morning, one can see the beautiful Florida Moonflower. This night-blooming Morning Glory (what a contradiction in terms!) is a truly lovely addition to our flora. Often three inches in diameter, these beautiful white flowers will close shortly, but this early on a cold morning they can still be seen in full bloom along the drive. A fascinating experiment is to find a just-before-opening Moonflower bud in the evening, about sundown . . . hold it in your hands and carefully observe it. In a very short time, the flower rapidly opens and displays its fresh beauty.

At station number four, a small footbridge crosses the borrow ditch to a grand vista of the wilderness that lies beyond. Near—and sometimes on—this bridge may be seen Mangrove Crabs which climb the trees and behave rather as do squirrels, hiding on the opposite side of the tree as you approach. The Mangrove Crab is one of the choice delicacies eaten by our two species of Night Herons—the Black-crowned and the Yellow-crowned. These birds can sometimes be heard crunching Mangrove Crabs much as kids munch popcorn at the movies.

All of the area visible, as far as the eye can see from the end of this walkway to Pine Island Sound, became "wilderness" under the United States Wilderness Act, on October 19, 1976 when President Ford signed a bill which included this part of Sanibel Island in the National Wilderness System, giving it complete protection in perpetuity.

Visible from this vantage point are hundreds of what ornithologists call "peeps." These are many species of small shore birds, and at this moment we're looking at about fifty Least Sandpipers only a few feet in front of us. In the distance are Snowy Egrets, Great Egrets, White Ibis, Louisiana Herons, Greater Yellowlegs and the Black-bellied Plover.

Evidence of Raccoon is everywhere—their tracks are on the dike road and as we look out across the flats, we spot two of them crossing from one mangrove island to another. There are quite a number of other mammals here. The Marsh Rabbit—a dark little fellow *sans* the familiar "cotton tail" — is abundant and on this drive.

The O'Brien Observation Tower offers a fine vantage point from which to view vast areas of the Refuge. On a good day, thousands of birds can be seen from this point. Near this tower one can often see a large number of Ring-bill Gulls, Laughing Gulls and occasional Herring Gulls resting on the road. Among them, often, are Black Skimmers which, after their morning skimming activities, like to rest and digest on the road in the warm sunshine.

The Otter is quite common just past the O'Brien Tower, where the road turns. This is sometimes called "Otter Corner" because their slides and tracks can frequently be seen on both sides of the road at this point. The sharp-eyed observer might see an Otter playing in the mangroves or diving for food in the borrow canal.

Alligators often crawl across from the left hand side of the road to the Bay to hunt at night and haul back again by morning. The alligator track is characterized by the dragged tail mark between the foot markings; usually if

Snowy Egret

you see a track going to the right you can find the corresponding more recent leftward sign.

Indigo snakes, which are very large, are sometimes seen gracefully gliding across the road. The Mangrove Water Snake is occasionally seen as is the Diamondback Terrapin.

This brief glimpse of the "Ding" Darling will perhaps encourage the reader to return here time and again. It is quite impossible to gain an appreciation of the life systems and the many species that occur here without spending many, many hours of careful observation and study.

Little Blue Herons. The juvenile is white.

The Anhinga, also known as the Water Turkey, Darter and Snake-bird, swims with great agility under water to capture its prey.

Yellow- crowned Night Heron

9 *A DAY IN THE LIFE OF THE SCCF WETLANDS SANCTUARY*

This 207-acre freshwater wetlands tract was acquired for the Sanibel-Captiva Conservation Foundation by dedicated conservationists at substantial expense, so that our rare and fast-diminishing barrier island freshwater system could be preserved at least in part. Many people were involved in this effort, but probably the most active were past Conservation Foundation chairmen Roy Bazire and Ann Winterbotham. Much credit goes to their foresight and energy in acquiring this tract.

Sanibel was thrown up by ancient seas as a series of somewhat parallel shell and quartz sand ridges. The two main ridges on Sanibel Island are the ones now followed by the Sanibel-Captiva highway which is the Mid-Island Ridge, and the Gulf Beach Ridge on the south. In between are quite a number of sets of minor ridges, some of them parallel, some of them criss-crossing each other. Lying between the two main ridges are the only remaining fresh-water wetlands extant on any South Florida barrier island. The islands of Palm Beach and Miami Beach had this resource at one time, but on both it has been destroyed by development.

Lying anywhere from 15 to 25 feet below the surface of the ground in the fresh-water zone between the two main ridges is a layer of dense sedimentary clay-like material. In theory, this layer prevents the salt water from encroaching from below and holds the fresh rain water as it falls on the basin thus formed. In reality, however, this system has been rather substantially violated in recent years by land development and particularly by the failure of the water control device at the Tarpon Bay Mosquito Control drainage exit. High spring tides allow massive salt water intrusion. Many marine forms can be found long distances inland in the "fresh" water area. However, it is still fresh enough to support about 35 species and subspecies of vertebrate animals that occur on our Sanctuary Island. The ability of some of these animals to adapt to saline conditions is quite remarkable.

To reach this tract, one travels from Tarpon Bay Road toward Captiva on the San-Cap Highway. Look for the new conservation Center on the left, where you will find a small parking lot. Come with me now as we make an imaginary trip together through this most interesting Sanctuary.

Leading from the parking lot is the main entrance to the Nature Trail and the Sanibel River wetlands. This entrance leads down an artificial low dike road which, in this writer's opinion, should never have been built, for it is completely unnatural and crosses the ridges and the swales alike, providing a barrier to the normal sheet flow of water in the rainy season. Moreover it is constructed of alien material trucked in from elsewhere on the Island.

As we walk down this road we note that it is called "Center Road." Soon we pass the junction of "Fern Trail." Pursuing our course in a generally southerly direction, on the left we pass the controlled-burn area, which was the subject of much discussion in the fall of 1976 when Foundation Director Richard Workman was able to convince his Board that a controlled burn would be a sensible experiment and demonstration.

What I call "the Smokey Bear Syndrome" has died (as has Smokey). Foresters and naturalists everywhere today are of the belief that burning certain kinds of natural habitat under special conditions, is quite the best way to manage it. In the case of Sanibel, there are vast areas of wild lands which have not burned in recent years and which today are covered with many, many tons of fuel—that is, dried grasses, dead branches, leaves and other combustible organic materials. Controlled burns during the wet season, with adequate precautions from fire protection agencies, seems to be the way to prevent disastrous dry season wild fires. Some controls used in this case were aircraft laden with water and a standby fire truck, together with experts from federal, state and local agencies, a number of whom were professional foresters, all of whom lent help and advice during the operation. It was a wet day; there were no serious problems and the burn was considered a success. But picture what would happen if, at the height of the dry season, a wildfire took off on a dry windy day across those fuel-burdened central wild lands of Sanibel. The potential for great destruction is scary. It is therefore felt by many authorities that controlled burns in the wet season should be repeated as needed and would cover, eventually, much of the Island. Such activities would not only enhance the environment — which we will discuss in a moment — but would also go a long way toward preventing the occurrence of dangerous wildfires.

How does a fire protect the environment? Well, for one thing, it helps in the destruction of encroaching exotic plants. True, our worst exotic, the Brazilian Pepper tree, *Schinus terebinthefolius,* was not seriously affected by the experimental fire, but other exotics were, such as the Australian "Pine," *Casuarina equisetifolia.* One can see dead ones at this site as a result of this controlled burn.

The predominant plant, *Spartina bakerii,* or Cord Grass, which is the indicator plant of our freshwater wetlands, came back in beautiful green clumps. One can easily see that if wildfire were ever to break out in the vicinity, this particular area would be safe.

As we walk down Center Road, we come to a junction called "Sabal Palm Trail," where we turn to the right. This takes us along the axis of a beautiful minor ridge.

Let me say here that the existence of posts and signs is not entirely acceptable to many members of the Foundation who feel that a wild land should indeed be wild and not cluttered with road signs. Actually, the "Ding" Darling Refuge has what many believe to be a superior system—simply numbered stations coded to its literature. This sign I'm looking at now says "Center Road," "Sabal Palm Trail" and "Lantana Trail." It all seems unnecessary.

We're turning down this minor ridge and walking past obviously burned Sabal Palms, White Indigo Berry and Guava. About 50 feet down the trail, we pause to look at a beautiful Cherokee Bean *(Erythrina herbacea).* This particular plant was donated by a local citizen who no longer needed it but didn't want it destroyed. It was moved into this site in January 1976 by some British Soldiers who were visiting here at that time. It's doing beautifully and is the only example on this tract.

As we proceed along the ridge we see that already this trail has been discovered by the public for we see scraps of tissues and beer cans. The latter, when found on the beach, have been termed "Pig Shells." It's unfortunate that Americans seem bent upon defiling even their few remaining wild places.

Also discernible here are old furrows and crop rows where, before World War I and until a hurricane wiped out agriculture in the twenties, cultivation of tomatoes, eggplant and other crops took place rather extensively. It is probable that the existence today of the Guava and Castor Bean can be attributed to this former activity in these "old fields" as the ecologists call them.

On this December day, I'm finding very few edible Guavas. Here's one that's not too bad. Eating around the Diptera larvae that are in it, I find it's quite tasty. We've passed other trees — the Sea-Grape tree, *Coccoloba uvifera,* familiar to most of us and much loved for its beauty and tasty fruit.

Did you know that the Sea-Grape leaf is useful? It may be used as a post card. That's a lot cheaper than buying cards in the shops and it's even more acceptable to the recipient. You can write on the leaf with a felt tip pen — or even a ballpoint will do. Many Sea-Grape leaves are exquisitely beautiful — rather flexible and leathery in texture. You can put a stamp right on the Sea-Grape leaf, address it, write your message on the other side, and this will go through the mails. However, I usually prefer to put my leaves in small envelopes. So choose a small leaf — then you won't spend your whole vacation writing lengthy post cards. Yet communication will be just as successful because your correspondent is guaranteed to pick up the phone and call you. You then say what needs to be said on the other person's nickle and the whole communication episode is entirely satisfactory.

Along the ridge, we just passed a little conical-shaped hole that has been rooted out by an Armadillo, of which there are many here. And there, high above our heads, hang the webs of the beautiful Golden Orb Weaver spiders. I just spotted an airplant, *Tillandsia* species, one of the eight kinds of bromeliad on Sanibel. Incidentally, the bromeliad family is the one to which the Pineapple belongs. *Tillandsias* are epiphytic, not parasitic; they get only physical support from the host plant — not nourishment. The familiar Spanish Moss is also a *Tillandsia.* It is one of the few that does not grow in the form of a rosette. The pineapple top is a typical bromeliad rosette.

All about us are crinkly, shiny-leafed plants, *Psychotria undata,* our Wild Coffee. A beverage that tastes something like coffee can be made from the beans of this plant.

The palms so far are all *Sabal palmetto,* our State tree. Even the little plants that look like blades of grass in the middle of the trail are seedling Sabal Palms. Notice how many of these palms have been burned, yet others right alongside seem not to have been. Several phenomena are at work here. One is that when the fire, which was caused by lightning, came in 1971, some trees were covered with loose petioles. When they burned and their basal portions eventually fell off, the tree was quite clean and showed no evidence of fire. In other cases, when the trunk was covered with leaves and longer petioles, the tree burned and you can still see burned petioles which, when they fall off some time in the future, will leave clean unburned trunks. In still other cases, dry leaves allowed for super hot fires that burned down to and completely charred the trunk. Of course there are all gradations of these conditions.

I've done a study of this fire. There's a lot to learn from the natural wildfire, but first you must learn its language. It's not possible in this short discussion to go into the subject in any detail, but from this fire one can get very accurate estimates of palm tree growth in several situations. For indeed there are some trees that were not touched at all, such as the great one in the middle of the trail that has every leaf it ever owned still hanging on it in a grandiose skirt rather in the fashion of the *Washingtonia* palms that you see also in California and Mexico.

Ah, I just spotted a Gopher Tortoise. He's heading very rapidly — for him — toward his oval-shaped hole that may be up to forty feet long. Almost entirely vegetarian, here he picnics on fallen Guavas.

Here is our Wild Cotton. Wild Cotton has been a federal boondoggle ever since FDR's New Deal time. It is, indeed, a cotton and as such it is host to cotton-destroying insects such as the Boll Weevil. But somebody—I think it was Henry Agard Wallace of the Roosevelt Administration — felt that two birds could be killed with one stone (if a conservationist may be permitted such a phrase), and he organized a Wild Cotton destruction program in the United States which continues to this day. This tract is full of Wild Cotton and every so often the Feds come in with their machetes and cut cotton. But they do it very carefully, in such a way as not to obliterate it completely. As a matter of fact, I can show you several plants here which have been carefully preserved — seemingly so that the jobs may also be carefully preserved. And this has been going on for decades!

The path is lined with Periwinkles, *Vinca rosea.* Did you know that this plant is originally from Madagascar? It is here to stay, however, having taken over so many of the hearts and lands of Sanibel. The Periwinkle is the source of an alkaloid called vincristine, a chemotherapeutic agent used successfully in the treatment of leukemia and other kinds of cancer. It seems to arrest mitotic division of some neoplastic cells at the metaphase stage.

Here's Leather Fern, the great six to ten-foot ferns that are all about us. The little curled circinate meristem, or "fiddlehead" bud, is edible. Don't pick them here, please, but if you find them on vacant lots, pick a dozen 6-inch examples and cook them as you would asparagus. You will find they're better than asparagus.

Many of the Cabbage Palms are covered with Poison Ivy. Not long ago I showed this ridge to 17 high school students and their teacher from a Louisiana girls school. All of them were just regular kids except one. Betty insisted on experimenting with Poison Ivy. She pulled over 250 pieces by wrapping the stems around her right wrist. Betty may become a great scientist if she lives! She promised to write, but so far no word. I hope her problem is nothing more than learning to write with her left hand.

Notice that some of the Cabbage Palms are littered at their bases with great heaps of petiole or leaf stem ends. This is a result of Raccoon activity. When Cabbage Palms are in fruit, there's quite a bit of competition among many animals who feed on the fruit. The coons rush up and struggle against each other to get to the fruit first and the loose petioles fall off — and sometimes the coons do too — hence this great heap of litter beneath the tree.

There are many other mammals here. Occasional Bobcats are seen. Our own Sanibel Rice Rat, which evolved on this Island and whose technical name bears the designation *sanibeli;* the Marsh Rabbit; Opossum; and a rather nice little rodent called the Insular Cotton Rat — which is quite a decent sort.

Deer used to roam this region, according to "Uncle Clarence" Rutland, who told me he saw 18 of them in a herd at the Lighthouse 65 years ago. I'm not sure but what there aren't some Deer here now. We had three "possibles" in 1976. Councilman Francis Bailey's gardener swears he saw one in the Bailey yard. A friend saw cloven hoof tracks on his property. And, finally, I was walking through this trail not long ago when I startled something which made a tremendous crashing effort to get away from me. I was unable to glimpse it, but I can't imagine what it could possibly have been other than a white-tailed Deer, unless some child was playing hooky from school. It's not inconceivable that with hunting pressures on the mainland, Deer would cross over here again, so I'm coming around to the belief that possibly we have a few. This will probably be proved all too soon when, with the ever-increasing amount of speeding traffic, one will undoubtedly get smashed on the highway, along with the other twenty-some animals that are killed every day on our Sanctuary Island.

We have just passed a nice little Olive tree, *Forestiera segregata,* which really does produce little olives. They're so small that if you used them in your martini, the cocktail would have to be served in a thimble.

Oh — there, hanging above the Olive Tree, is a Crab Spider. The orb-weaving Crab Spider, *Gasteracanth* sp., actually has a carapace like that of the Blue Crab, with points on the end. It can be picked up, if one is careful, between the thumb and forefinger, balancing the creature between its points. In that postiton, it can't reach you with its mouth parts or its legs. It's one of our finer orb-weavers, along with the great Golden Orb-Weaver, *Nephila clavipes.*

Here's a beautiful leafy plant called Myrsine. It used to provide a lot of good food for migrant birds as did the Marlberry, but both of these have been largely destroyed or crowded out by the ever-encroaching Brazilian Pepper. Here's a patch of Snowberry which, at this time of year, is unmistakable because of its — you guessed it — snow white berries.

There goes another Gopher Tortoise, noisily crunching through the dry vegetation. This is the first time I've seen two Gophers out here in the morning. This one is heading straight for his hole and is not the least bit interested in socializing.

I have seen other reptiles here, such as the Great Indigo Snake, the Ringneck Snake and the huge Eastern Diamondback Rattlesnake, to mention a few. None of these creatures offers any threat to a careful visitor. Should you encounter venomous snakes, prudence is required. But let the snake go his way and you go yours, and you both will enjoy the rest of the day.

Here's the great Gumbo Limbo Tree which I've named the Jumbo Gumbo. It is the largest specimen on Sanibel, I believe — a magnificent creation. You can't miss it. It's on the left hand side of the trail, right near several felled Cabbage Palm logs.

I imagine that when one first hears the name, Gumbo Limbo, it must be puzzling. Should you decline politely saying, "Sorry, at my age I no longer dance"? Or should you say, "You are very kind, but usually I drink Scotch"? It may come as a shock to learn that the Gumbo Limbo is our most treasured tree. It belongs to a family widely dispersed throughout tropical America.

A friend tells me that when she first came to Sanibel, she thought that the Gumbo Limbo must have had a religious significance for the early Indian settlers, since they were the prominent vegetation forms found growing on top of many Island Indian Mounds. Years later she come to realize that the reason the Gumbo Limbo grew there was due to its capacity to grow just about anywhere in well drained situations, under the worst kinds of conditions. Quite probably, the Gumbo Limbo was the only large tree hardy enough to propagate itself on those heaps of shells, bones and garbage.

Cabbage Palm, the Florida State Tree. Note the Poison Ivy and Virginia Creeper. The Racoons feed on palm fruit; in the competition, many old petioles get knocked down and pile up under the tree. A Leather Fern may also be seen.

Do you catch that slight skunk odor here? It's due to one of the Stoppers — *Eugenia* species. Many people feel that they've smelled a skunk when they walk through this area. It is possible, since two species of four-legged skunks are recorded on Sanibel, but they are very rare. Usually when this odor is detected, it's *Eugenia.* Early folk medicine employed Stoppers to control infant diarrhea.

Also near the Jumbo Gumbo Limbo are several wild Papayas. Papaya, you may know, is the source of papain, a protease — that is, a protein-digesting enzyme. It is from this plant that commercial meat tenderizer is extracted. Primitive peoples used to wrap their tough meat in the leaves of this plant, put it out in the sun and it would become tender. (I sometimes feel that if you put a tough old steak out in the sun wrapped in the *Wall Street Journal* it would work just as well!)

Less primitive people sprinkle this enzyme on tough steaks as we do here in America. There is a school of medicine that injects this substance into the slipped disc of the lumbar vertebral region on the theory that the ensuing protein digestion will relieve the pain by softening the disc. This is a very controversial procedure that is accepted by some and is scorned by others. Some doctors of my acquaintance are in the latter category.

Our tiny much-hated "no-see-um", or sandfly, tortures us with his bite, depositing an alien protein in our skin. Papain, in a weak aqueous solution — and this can be made from a little bit of Adolph's Meat Tenderizer — can be rubbed onto such irritated areas of your skin and almost immediate relief is experienced by most people. This is a folk remedy used widely in Florida and is not original with me. (FDA please note.)

Near the Jumbo Gumbo is a Strangler Fig, *Ficus aurea.* It displays the "Banyan" habit of many members of this great genus of more than six hundred plants worldwide. The Strangler Fig can be readily identified by the reticulated mass of adventitious roots that hangs down from this tree. A better specimen is on the far side of this tract, about a mile from here if you walk the perimeter route. The seed of the Strangler Fig is dropped by a bird into the axil of a Cabbage Palm, sometimes high up, sometimes near to the ground. It germinates in that same axillary region, sends its roots to the ground, continues to grow and eventually envelops the Cabbage Palm. I have seen large free-standing fig trees with only a vestige of a green palm leaf protruding from the top, but normally the two species exist together quite successfully for long periods. Some excellent examples of this coexistence can be seen in the vacant area just a few feet from the Sanibel Elementary School.

A few months ago, near the Jumbo Gumbo, I discovered what I thought were a series of fox holes. We do have a few reliable Grey Fox sightings on the Island, and I believe I was correct in my identification of these holes. But now they've been abandoned and are hard to find since much palm litter has fallen on them. As you walk these trails, you never know what you're going to see next.

If you look about you, you will see round holes in the ground here and there throughout this ridge trail. The Cabbage Palm is about the only tree I know of around here that, when it dies, leaves a hole in the ground rather than a stump sticking up. These holes are the result of the soft vascular tissue in the center part of this "monocot" plant, which decays first and drops straight down to a foot or more below the level of the ground after the tree dies. The harder peripheral or cortical material retains its shape, but eventually, after a year or so, when it finally falls, there is this neat round foot-deep hole in the ground which is quite surprising indeed.

The tiny inverted conical depressions in the ground under dry parts of leaning protective trees or leaves, are Doodle Bug or Ant Lion traps. These traps are occupied at the bottom by a Neuropteran larva which flips the sand out in this absolutely symmetrical fashion. The sand rests at the angle of repose which for this material is about 26^0. The beastie stays at the bottom and waits for an ant or other small insect to become entrapped in the loose sand. When that happens, a lot of activity ensues as the animal flips its mandibles, tossing more sand up, causing the ant to fall on miniature landslides to the bottom point where it is grasped and eaten. After about six sheddings of its skin, this larva turns into a rather beautiful gossamer-winged creature not unlike a dragonfly.

About 250 feet past the Jumbo Gumbo, on the right hand side, is a broken down Buckthorn Tree. I don't know how it got broken this way — probably some storm years ago — but it's still alive and is retained here because it is a natural casualty and because, as you can see, it is a fine host to many epiphytic bromeliads, of which there are two species almost entirely covering it.

Cat's Claw is the next tree on the right. During some seasons of the year its young leaves are as beautiful as flowers, but watch out — the Cat's Claw is well named. It has thorns that will get you if you don't watch out.

Be careful too, of the Spanish Bayonet or *Yucca* which you must pass next. Please don't touch it because those sharp spines are almost impossible to touch without experiencing severe pain. The flower parts of this plant are beautiful white succulent things and are delicious, as noted in Chapter Seventeen.

Right here I would like to comment that some of the management practices of this tract are beyond my comprehension. Just behind the Yuccas is a badly damaged young Cabbage Palm, which has obviously been intentionally cut. So often I see valuable trees that have been destroyed, yet the weed trees, such as the Brazilian Pepper, are left intact. I have hopes that the Directors of the Sanibel-Captiva Conservation Foundation will address this deplorable situation head-on and do something about correcting it.

Also by the Yuccas is a small herbaceous plant called *Mentzelia floridana,* commonly known as "Poor Man's Patch." This plant has little hooks all over its leaf, rather like "Velcro," and a piece of this leaf could be put over a hole in fabric and make a satisfactory, if temporary, patch. When it's in bloom you can call it "Instant Corsage," because you can pick a few of the beautiful little yellow flowers and they will stick immediately onto a fabric.

Soon we come to another species of palm, on the right side, which, unlike the Sabal Palm, has a rough serrated petiole. This is the Scrub or Saw Palmetto, *Serenoa repens,* which is so common throughout Florida and is an indicator of Rattlesnake habitat. There are very few Saw Palmettos in this tract but the ones that are here are rather unusual. Unlike most of this species in Florida, these have reached for the sky and some are 17 feet tall. In most other places, they are short, stocky plants, with most of the trunk buried underground, as a sort of fire protection adaptation.

Another hundred yards farther along, on the left, is the great bee tree. This old dead Strangler Fig will fall one day, and I don't want to be near when that happens because those bees are not going to be very happy. This is rather an unusual beehive because it is not inside a hollow tree, but hangs on the outside of the tree, really exposed. With field glasses you can examine the life of the bees at a distance of about 40 feet. In the winter time, this hive population just about expires, perhaps due to weakening from summertime insecticide sprays, and in fact I've seen massive bee kills from the mosquito planes that spray the Island in the summer. In the winter of 1975-76, we felt that the queen must have died because the population became smaller and smaller every day. A queen was purchased, but she arrived very late and the hive had already recovered. Obviously the original queen had not died or, perhaps, the workers had another viable queen cell. Whatever the explanation, the hive recovered very well during the summer of 1976, but during the cold weather of 1976-77 it again showed considerable decline. I started to feed the bees that winter with candy and honey. It seemed to help and as of this writing (mid-1977) the colony seems to be recovering again.

A few feet beyond the bee tree is what is probably the only *natural* body of permanent fresh water on Sanibel Island. This is an ancient Alligator hole. We know this because the dorsal osteoderms of a very large Alligator have been taken from the mud bottom. The Alligator in primordial Florida used to provide a very very useful service to all of the other animals of the region, by providing water in the dry season. These Alligator holes never dry up as they are always below the lowest water table. All of Sanibel's canals, real estate lakes, golf course water ponds, whatever — all of these are unnatural, and to my knowledge, this old Gator hole is the only true natural body of permanent fresh water on Sanibel.

That is not to say that the slough system in ancient times on this Island did not have water at certain seasons, but it was not always present because it tended to dry up in the dry season.

All of the trails on this tract add up to something around five miles. I computed that in 1975-76 I walked over three hundred fifty miles on this one two hundred seven acre tract, so I guess I know the place about as well as anyone. One of the beautiful things that happens here happens one night each month — the night of the full moon. If it is a clear night, I bring friends on a walk through these trails. No flashlights are allowed. No one talks and there is no disturbance whatsoever. We creep through these trails to see what we can see by the light of the moon only. It's an eerie magical experience to walk through this lovely ridge trail at night without artificial lights.

Another activity that I undertake on these trails is the study of mini-habitats, as I call them. They consist of particular fronds or fallen trees that are useful hiding places for many of the creatures that live here. I have identified about ten of these and usually, in walking through this tract, I can turn up snakes and frogs and sometimes interesting invertebrates such as the Vinegaroon, *Mastigoproctus giganteus,* which, if you hold it in your fingers, makes your hand smell like salad dressing. This is also known as the Whip Scorpion which, although horrible to look upon, is absolutely harmless.

Here and there on the ridge you will come across seashells. Obviously these are thousands of years old — they've been here ever since the ridge was thrown up by the sea. So if you're interested in old, really old, shells, take a good look at these.

Just before you come to the Bee Tree, there's a left turn which takes you to the slough or swale that has paralleled our walk down this ridge. As you will observe, it is almost entirely clean and free of exotics and is, therefore, a fine swale indeed. Its principal vegetation is Cord Grass, *Spartina bakerii,* the indicator grass of the wetlands. That particular slough, however, is not entirely natural. I am told that about a foot of topsoil was removed from this slough a few years back. This compensated to some degree for the artificially lowered water table, and this swale, the most beautiful on this whole tract, was the result. Similar practices might be successfully used in the restoration of swales elsewhere on the Island. And there's a by-product, too. Such soil can be quite rich and would be useful in horticulture — a possible source of income for the Foundation.

There are many many other animals and plants and other trails that we could explore in this discussion, but we've got to leave something for you to discover for yourself. Let me suggest that you make your own visit to this tract. Treat it gently — not like the motorcyclist I encountered who ran his cycle over a Gopher Tortoise and killed it, something that should never happen and especially not in a Sanctuary. And please don't smoke, because wildfire can be very serious.

Help us take care of this natural treasure and if you've got any good management ideas, why not pass them on to the Sanibel-Captiva Conservation Foundation? Visit the new Conservation Center and discuss your questions and suggestions with the trained personnel there.

10 *FISH CONTROL THE PESKY MOSQUITO*

The most important fish around Sanibel are not Redfish, Snook or Trout, but tiny little things that most of us ignore.

Gambusia affinis, the Mosquito Fish, is perhaps the most abundant fish in Sanibel waters. This species is native here, ranging from Texas all along the Gulf and up the Atlantic as far as New Jersey. By some authorities it is divided into two subspecies, the typical form and the subspecies *holbrooki.* I find these two subspecies almost indistinguishable, hence this discussion will simply deal with the species as a whole.

The word *Gambusia* means "worthless" or "nothing" (new Latin). Probably some early aquarist compared *Gambusia* with its more beautiful relatives such as the Guppy, Platy or Swordtail and felt it to be worthless as an aquarium fish. But he overlooked the animal's immense value as a predator of mosquito larvae and as a very important item in the complex food web of a place like Sanibel.

Gambusia belongs to a group called Tooth Carps which occur naturally only in the New World, although many have been introduced throughout the world for mosquito control purposes. *Gambusia,* in common with numerous Tooth Carps, is a live-bearing fish. The eggs are fertilized within the ovarian cavity, but their nourishment is derived solely from the egg yolk material. This reproductive process is called ovoviviparity.

Some Tooth Carps are truly viviparous, because in such cases there is actual nourishment of the embryo derived from the mother. Still others are strictly oviparous in that they are egg-laying Tooth Carps. We will touch on some of these later in this discussion.

In *Gambusia,* the male is equipped with a copulatory organ known as the gonopodium, which is derived from the third through fifth rays of the anal fin. It forms a kind of channel and serves to transfer sperm to the female's vent.

As can be seen from Molly Eckler Brown's accompanying drawing, the female *Gambusia,* at approximately two inches in length, is substantially larger than the inch-long male. Note the dark spot on the female, known as the pregnancy mark. Note also the gonopodium of the male. Observe that one male is black speckled. A certain small percentage of males, mostly those living in fresh water on Sanibel, have this black speckled characteristic. It is possible to selectively breed *Gambusia* to solid black in the aquarium, and more than one experiment has produced an entirely melanistic race.

Many Tooth Carps, including *Gambusia,* are able to tolerate either the fresh or the salt water habitat. This of course is very valuable to us because this species can prey on fresh water or salt marsh mosquito larvae. Space will not permit a discussion of the physiological changes that a fish must undergo when traveling from fresh to salt or from salt to fresh water. Suffice it to say that it is a tremendously complex matter and one marvels at the adaptive ability of such an animal.

A large female *Gambusia* may produce forty young at one birth. The pregnancy period is about three weeks, after which there is approximately a week's rest period before she becomes pregnant again. Thus one female may produce almost 500 young per year. And if there doesn't happen to be a male around, it doesn't matter as she may well have stored a supply of viable sperm from the last time she was bred.

Many of the young will achieve sexual maturity during the year of birth and consequently it has been computed that more than 15,000 young could theoretically be derived from one pair of *Gambusia* in one year.

Before we leave the sex life of the Tooth Carp, we note two other interesting details. Some species are the Christine Jorgensens of the fish world, in reverse. A productive female, a proved mother, may and often does fully transform into a functional male.

Another peculiar Tooth Carp characteristic is that some species exist as entirely female populations. This is true of the "Amazon Molly" which derives its name not from the great river, but from its totally female population. This Texas species mates with males of related but *different* species and thus pregnancy results, but the "father" does not contribute anything to the heredity of the offspring. And all offspring are females. He only "triggers" the pregnancy, without adding any genetic material.

Of course in addition to being predators the *Gambusia* are also singularly important as a prey species. On the "fresh water" side of the Darling Memorial Drive there are sometimes thousands of birds. In addition to crustacea and other invertebrates which are being preyed upon by these birds, and in addition to a few other fish species, the *Gambusia* is perhaps the single most important item on their menu.

You have perhaps climbed down the bank of the Wildlife Drive and witnessed a flurry of surface activity as a shadowy wave rushes from the shore when you appear. This would probably be a school of *Gambusia*. Almost their total activity is conducted at the top of the water where they seek air-breathing insects and some plant materials. Hence the surface flurry as you approach. And hence the infinite patience of the Green Heron which often waits for the *Gambusia* to appear below his particular mangrove root. Hence also the fortune of the Black Skimmer which with considerable skill and grace has great success at harvesting his share of *Gambusia* from the surface of the water as he flies low, knifing its surface with his sensitive lower mandible. When contact is achieved, it's snap, swallow and poise to skim again.

It is perhaps of interest to note that the "Ding" Darling National Wildlife Refuge Memorial Drive was not constructed for the purpose of viewing birds but was in fact designed to impound water for mosquito control.

The impounded water of the left-hand side of this dike road serves two purposes. First, it reduces the breeding environment of the Salt Marsh Mosquito, *Aedes taeniorhynchus*, which must oviposit on exposed land. As an adaptation to seasonal water supply, the insect lays its eggs on land that will, in the natural course of events, become flooded either by rain or tide. The strategem of mosquito control is to maintain the land flooded at all times, thus reducing the egg laying habitat of the mosquito.

But a second and equally important reason for the construction of this dike, and for that matter, the mosquito ditches found throughout Sanibel, is to maintain permanent aquatic habitat for larvivorous or mosquito eating fishes. So this two-pronged assault against the Salt Marsh Mosquito has worked very well and, as a bonus, resulted in a fine wildlife drive — one of the finest in the world.

The mosquito control activities also provide a wonderful, though artificial, habitat for wildlife. Mosquito ditches have allowed for the expansion of the Otter population. The impounded waters have created a habitat for numerous birds. Studies have been made that show some species of birds actually prefer the impounded, although unnatural, habitat created by man to the natural salt marsh conditions. This same study did not reveal the existence of a single avian species that was adversely affected.

Now let's look at *Gambusia* as a predator. A large female *Gambusia* weighs about a gram and consumes approximately her weight in food each day. This food consists largely of mosquito larvae when they are available. Studies have determined that a gram of mosquito larvae of a species that occurs in Singapore contains 890 individuals. Assuming that mosquito larvae here weigh approximately the same, regardless of the species involved, it's safe to say that a large *Gambusia* on Sanibel would consume approximately an equal number of larvae.

We have already noted that a mother *Gambusia* is a remarkably efficient engine of reproduction. So now if we attempt to compute what she and her progeny can consume in one year, we begin to see emerging a measure of the value of this animal in the environment.

At some seasons of the year, often in the cold months, we note that the *Gambusia* are thin, underfed and unhealthy. To cope with such slack food seasons, *Gambusia* has another interesting adaptation for survival and that is the consumption of its own kind, both young live ones and old dead ones. To some, this may appear counter-productive but on reflection one sees that this does provide an adequate and entirely successful method of species perpetuation and also, of course, provides for the best selection for the future gene pool.

When the rains come and the remaining salt marsh mosquito habitat that has not been impounded produces its first great bloom of mosquitoes there is a lag in predator action. A lag in predation at the onset of prey eruption is very obviously a logical happening. But as food becomes abundant the *Gambusia* set to work on the production of their own kind and after twenty-one to twenty-eight days a huge population increase is set in motion, and predation as a method of mosquito control becomes much more efficient. Some people tend to credit this success to chemical toxicants, but I'm sure that a bit of reflection will demonstrate that such a lag is only a function of the reproduction of these predatory animals.

There are other interesting predator fishes in our Sanibel system. One, *Fundulus confluentas,* can be termed to be an "annual". I have been observing one area on Sanibel which is not connected to the Sanibel water system in any way and which completely dries up during the dry season. However, shortly after the rains come and the area becomes flooded, *Fundulus confluentas* appears in numbers.

I suspected when I first observed this phenomenon that viable *Fundulus* eggs existed in the dessicated mud, somewhat analagous to the breeding habit of the salt marsh mosquito, waiting for the rains to come. This suspicion was confirmed after a visit of Dr. Thomas Lodge, a young Florida ichthyologist who stated that this species is in fact an annual. So in areas where there is not permanent impoundment of water, some protection against the onslaught of the despised insects can be achieved by employing this fish.

There are in the world some 28 annual Tooth Carp species and it might be worthwhile to investigate them all to determine whether or not one of them would do a better job for Sanibel than *Fundulus confluentas.*

***Gambusia** — a larger female and a partially melanistic smaller male — at home in a pond with Marine Naiad, **Najas marina,** a cosmopolitan salt-tolerant water plant very common on Sanibel. Marine Naiad is much used by ducks, coots and gallinules. All parts are consumed. Note the raft of mosquito eggs and larvae.*

Another little live-bearing mosquito fish that occurs in great numbers here on Sanibel is called the Dwarf Topminnow, *Heterandria formosa.* The adult female is less than an inch long and the male is much smaller. Babies of this form are about the size of a comma. For many years the zoologists of the world considered this to be the smallest of all vertebrate animals. But now, I understand, a minute frog species has been judged to be slightly smaller.

The familiar "Sailfin Molly" is known widely as *Mollienesia latipinna.* This is often seen in aquarium shops all over America and is closely related to, or may even have contributed to the genetic complex of, the "Black Molly" of commerce. Mollies, in addition to eating insect larvae, feed on algae and other vegetation. The fact that they do not ever eat their own young is a logical correlation of this for bad times for *Gambusia* are not necessarily bad for the Molly since algae of some kind are usually always present.

There are in addition to the above about five other mosquito-eating species that are active here but the foregoing discussion touches on the most important and interesting ones.

A fascinating activity that one can undertake is to gather a few of these small fishes and keep them in an aquarium or even a large jar. They are very hardy as they are adapted to living in polluted water low in oxygen. However if you use Sanibel tap water you must let it age for a day else it will kill your specimens. You will have an interesting experience watching life processes and identifying the several species.

White Ibis

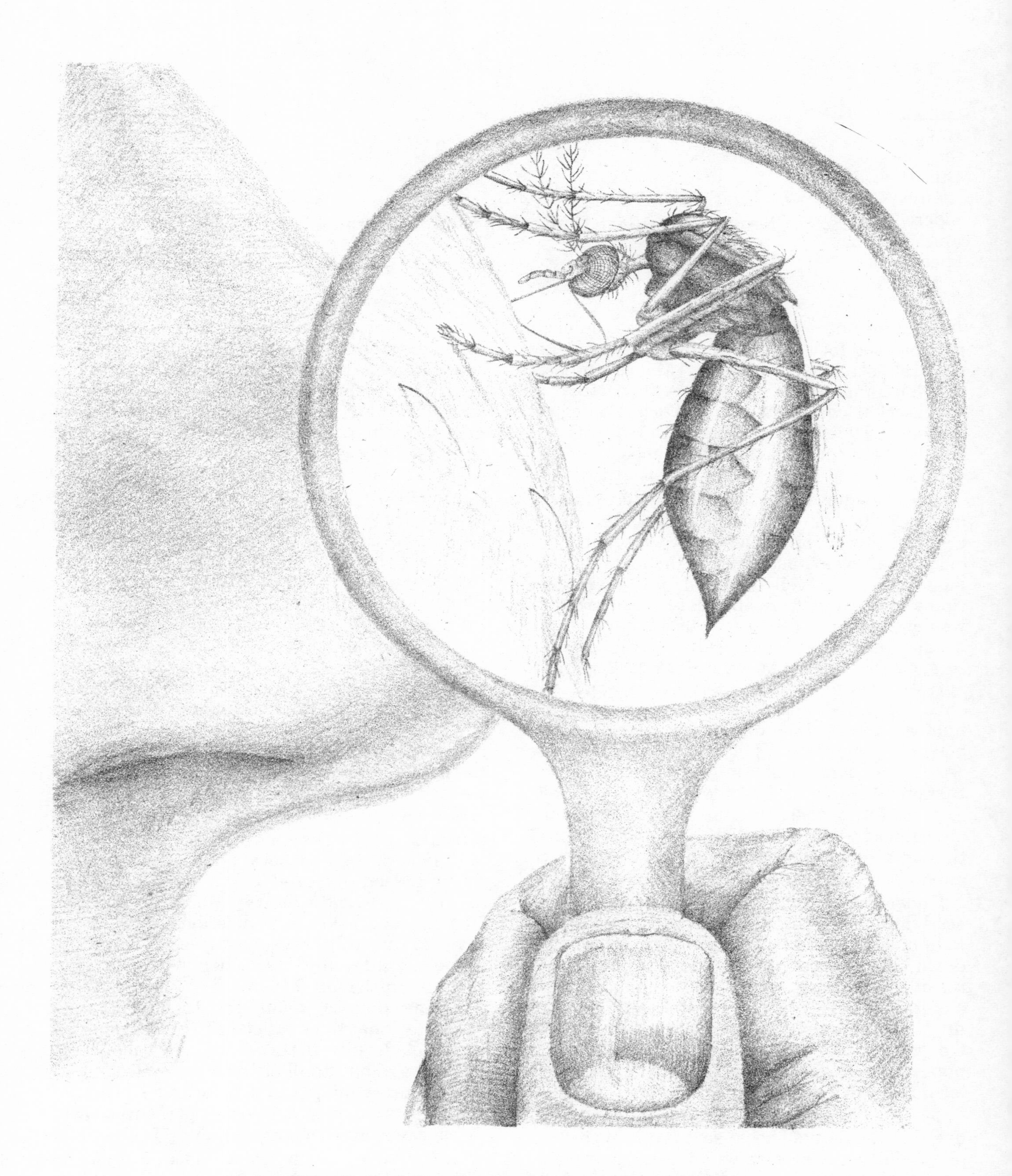

The author takes a close-up view of his subject.

11 *INSULT IN EDEN*
Chemical Mosquito Control on Sanibel Island

Some four decades ago when still a student, I participated for a summer in a study of fish life at Palmetto Key, where Mary Roberts Rinehart maintained a marine laboratory in cooperation with the New York Zoological Society. I assisted Marshall B. ("Doc") Bishop, who conducted the field aspects of this program.

The mosquito situation on that island, known today as Cabbage Key, as well as on Captiva and Sanibel, where we did much of our work, was absolutely unbelievable. Our laboratory was well screened, but at certain times when there were abundant mosquitoes, it was quite impossible to see out of the windows because the mosquitoes were so thick on the screens that visibility was blocked. To stand outside was often extremely unpleasant. We once computed that there were more than six hundred mosquitoes on the bare torso of one person at one time. That person was I. Being the junior member of the group, I was "Doc" Bishop's "Gopher". I would "go for" the bait, "go for" the beer, and in general "go for" anything anybody needed. So it was natural that one day, when Doc's interest turned to mosquitoes, he would select me to be the bait.

Under such conditions we did our work, studying, collecting, tagging, and releasing fish, mostly the already-fabled Tarpon, up and down this chain of barrier islands. The other residents of these islands also managed to get along, for there were, even then, several hundred hardy souls who lived and worked here. On the whole, and in spite of this super population of mosquitoes, I much prefer these islands as they were to what they are today.

Several years ago, I returned to Sanibel to live. I chose Sanibel because it is one of the last places in South Florida to have large areas of privacy, with some semblance of the kinds of wild places I knew in my youth.

I was shocked, one spring morning in 1974, to observe what seemed to be an air raid attack over Sanibel. On investigation, it proved to be a flight of ancient DC-3's, flying at tree top levels — and I mean about eighty-five to ninety-five feet altitude. They were casting an abundant fog of insecticide all over the landscape. They made repeated passes up and down the long axis of the island, much as though they were mowing a great lawn, and soon pretty much of the whole Island was blanketed by what I later learned was a fog of Baytex, or fenthion, a very toxic organophosphate pesticide.

Shortly after that, I observed helicopters applying another insecticide, Abate, also an organophosphate. Then, a few days later, while trying to reach my home just after sundown, I encountered what at first I thought was smoke from a huge fire, but which was, in fact, nothing but a ground-fogging vehicle, spewing forth a superabundant quantity of what I soon learned was Malathion.

My immediate reaction was that such a brutal insult to the environment of an island that is supposed to be a sanctuary was very wrong indeed. I visited T. Wayne Miller, director of the Lee County Mosquito Control District, and asked my questions, and I must say that Mr. Miller was most free in answering. However, after my discussion with that personable gentleman, I found that I still had just as many doubts about the technology employed by that very efficient organization and was still deeply disturbed at the callous onslaught against Sanibel Island's environment. In the intervening years of study since that date, no evidence has been presented that would cause me to change my opinion.

I am not unfamiliar with mosquito control activities, having spent several years in South America and Trinidad as the superintendent of a large *Anopheles* mosquito control program. *Anopheles* mosquitoes are serious business for, as everyone knows, many of them are vectors of malaria. Here on Sanibel we have some *Anopheles,* but no malaria. The major mosquito problem here is one of annoyance caused by Salt-Marsh Mosquitoes, mostly *Aedes taeniorhynchus,* which does not vector disease here.

On Sanibel Island, the areas set aside for the preservation of natural habitats — that is, the "Ding" Darling Refuge, the Sanibel-Captiva Conservation Foundation's land, other lands owned by The Nature Conservancy, some State of Florida lands, plus some private preserves, total about forty-five percent of the entire area of the Island. The rest of the Island is in private hands and subject to various degrees of development intensity, as specified in the Sanibel Comprehensive Land Use Plan.

Sanibel has attracted many people who don't really have a strong interest in conservation. Some of our citizens don't care a bit about the natural attributes of the Island; they're largely concerned with profits and business investments. Other residents realize the great importance of the environment and will do anything — anything, that is, short of limiting chemical mosquito control — to achieve environmental health. There are also quite a number of persons who are truly concerned with the environment and recognize that mosquito control employing quantities of lethal toxicants on this Sanctuary Island constitutes an intolerable environmental atrocity.

A number of zoologists have surveyed the wildlife on Sanibel and feel that there has been appreciable damage due to use of insecticides. This damage is discernible among many vertebrates, including reptiles and amphibians, as well as among birds. For example, a leading ornithologist commented to me on the diminished populations of songbirds in recent times. I once witnessed a massive bee kill. Dragonflies, fish, and especially crustaceans such as shrimp are also killed.

On August 29, 1975, Sanibel was enjoying beautiful sunny weather, the kind of summer weather that local residents love. There were literally millions of graceful dragonflies shimmering in the sunshine, all competing for the few flying insects present. Mosquitoes were not a problem, for these dragonflies, voracious predators of small flying insects, had virtually wiped out all of the adult blood seekers. On August 30, the "Lee County Air Force" came over shortly after dawn with their fogging DC-3's and spewed Baytex densely all over the Island. Four to five hours later, dragonflies were lying about the ground, disoriented, kicking, unable to fly — displaying the characteristic symptoms of cholinesterase depression. Soon all were dead, but not before Gray Kingbirds feasted bountifully on the dying predators.

The next day I found dead Kingbirds. A week later a new generation of mosquitoes was out in great numbers. On September 12 the planes flew again and more Baytex was spewed throughout the Island. It did no harm to the dragonflies this time, however, for all were already wiped out.

Such callous unthinking abuse is hard to tolerate. At this point it seemed very worthwhile to look further into the whole matter. There ensued a period of intense study and information-gathering. Fortunately I was able to secure the use of a great national laboratory's computer, and elaborate searches were made that produced literally bushels of information on the subjects of Baytex, Abate, Malathion, and, generally, on organophosphates and other pesticides of all kinds. Correspondence and discussion with officials of the United States Department of the Interior (USDI) and the Environmental Protection Agency (EPA) contributed valuable information. I attended a meeting of the Florida Anti-Mosquito Association (FAMA). This group held technical discussions during several days and I learned a lot about the attitudes and techniques of Florida insect-control professionals. I was able to assess Lee County's Mosquito Control activity in relation to similar activities elsewhere in the state.

Here are some of the things that have been revealed.

MALATHION

No longer do road vehicles spew the blinding fog that was a traffic hazard a few years ago. More sophisticated ultra-low-volume (ULV) application equipment is now employed. It throws out an almost invisible mist of fine matter that is no longer a problem to traffic and not much of a problem to mosquitoes. The use of Malathion still goes on but with decreasing efficacy because much of the mosquito population has now developed genetic resistance — that is, some mosquitoes are no longer killed by Malathion.

The use of Malathion on Sanibel today is, at best, a waste of money, the effects being largely psychological. It makes people *feel* that something is being done for their comfort. At worst, it is a public health menace, for Malathion itself can *cause* disease in human beings.

In 1975, St. Louis Encephalitis (SLE) occurred in about eleven states in the Mississippi Valley. Throughout almost that whole area, Malathion was the insecticide of choice and was employed to control those mosquito species that were thought to be vectoring SLE. About a decade and a half ago, SLE was a very real problem in the Tampa Bay area. And in 1977 an ominous few cases were reported in Lee County.

Should there be an epidemic of SLE on Sanibel, Malathion would probably not be chosen, for this widely-used weapon in the public health officer's armamentarium no longer works very well against some mosquitoes because of resistance. The "shotgun" application here has rendered Malathion at least partially ineffective. Some other, perhaps even more toxic, agent would no doubt be chosen.

BAYTEX (fenthion)

This is the product applied by aircraft as an adulticide. The United States Department of the Interior Restricted List includes fenthion, stating that it may be used only when non-chemical techniques have been considered and found inadequate and when use can be limited to a *small-scale application.* The USDI once informed me that they have no jurisdiction over what pesticides are used on Sanibel outside of the J.N. "Ding" Darling National Wildlife Refuge lands. They stated that they cannot regulate wind-drift from adjacent lands where they have no jurisdiction.

I was not satisfied with this response and took the matter directly to Nathaniel P. Reed, then Assistant Secretary of Interior for Fish and Wildlife in the Ford Administration. He agreed with my position and wrote the following to the Director of Fish and Wildlife:

"I do not believe the Service is aggressively pursuing this matter [of fenthion] in a manner that brings credit to the Service as a protector of fish and wildlife resources. Dr. Stickle's statement that 'in normal careful use in mosquito control, fenthion (Baytex) remains active in dead and struggling invertebrates long enough to kill at least a few birds, often shore birds, in nearly any application that is carefully studied' suggests to me that the scientific evidence on the toxicity of Baytex is sufficient to warrant its use only in the absence of suitable substitutes. Why isn't the Service moving more aggressively to restrict mosquito control to the use of less toxic pesticides in and around the Refuge environs?"

Similiarly, I took my case to Russell E. Train, who was top man in the United States Environmental Protection Agency (EPA) under President Ford. I had discussed these matters on the phone and in correspondence with EPA's Southeast Regional Office in Georgia for a long time and had arrived nowhere until I took the case personally to Mr. Train. Response was immediate. Both EPA and USDI initiated studies which are "ongoing", as they say in government bureaucratese, today on Sanibel as I write these words (mid-1977). Both Mr. Reed and Mr. Train are no longer in office, but the mill wheels that they set to turning are still grinding out very valuable data on the misuse of lethal toxicants on Sanibel Island.

The first efforts in response to Mr. Reed's initial memorandum quoted above proved not acceptable to him. Assistant Director of Fish and Wildlife George W. Milias wrote a memorandum to Mr. Reed on August 16, 1976. It was a typical bureaucratic two-page memorandum stating that drift onto their property cannot be controlled by the Fish and Wildlife Service and that their agreement to permit "incidental" application of Baytex on some Refuge areas was a responsible and defensible action.

Assistant Secretary Reed replied on September 28: "Deputy Director Milias' memorandum of August 16 does not constitute a satisfactory response concerning pesticide use on Sanibel Island, Florida, and I expect a much more thorough review of this matter to be accomplished in an expeditious manner." He discussed various aspects of the mosquito control program and ended his list of items with: "I cannot believe that there is no safer alternative to the present program. To clarify this issue I specifically want Dr. Stickle to personally review this entire matter including talking with Dr. Hicks at EPA. A full report is to be prepared for me which discusses the knowns, unknowns, and recommended courses of action including added studies, if necessary. The Service can and should do a better job than this on such an important matter — to do less is to seriously neglect the Agency's mandated responsibility concerning Refuge management to protect fish and wildlife."

The Dr. Stickle he referred to is William H. Stickle, trained as a herpetologist (student of reptiles and amphibians) and product of the University of Michigan, a scientist with whom, some 40 years ago, I collaborated on herpetological collection, observation, and studies. Today he is at Patuxent Wildlife Research Center and is possibly the country's leading authority on pesticide effects on wildlife. Dr. Stickle's report to the Assistant Secretary is dated October 27, 1976. It has this to say about fenthion-Baytex:

[Fenthion] "is notorious for having an especially high toxicity for crustaceans and birds. Because of its hazards to crustaceans such as shrimp, crabs, and waterfleas, some mosquito experts do not wish to use it. Damage to birds is expected only at levels above those that threaten crustaceans. My published comment on fenthion in relation to birds to which Mr. Campbell refers, is from various studies in which fenthion solutions were sprayed directly on marsh areas at 0.05 or 0.1 lb/A (pounds per acre) active ingredient. Under these conditions a limited amount of mortality or illness of birds was detected by careful searchers. At Sanibel, where aerial fog is used, deposition is far less, about 0.002 to 0.005 lb/A, according to Mr. Hicks. I know of no work in which the effects of such applications on birds has been examined. One would not expect visible effects, but some birds might be harmed by feeding on large numbers of poisoned insects."

Photo by the author

Air raid

He goes on to say under a title: "Known effects of fenthion on Sanibel":

"Mr. Hicks states that in fenthion-treated areas on Sanibel the fog results in definite kills of pink shrimp, even when live cars holding the test animals are below the surface or under cover of a wharf. This is supported by laboratory tests. The Gulf Breeze Laboratory found that 30 to 40 parts per *trillion*[1] killed young shrimp. Hicks found that 66 parts per trillion gave 100% kill.

"Yet at the recorded rates of deposition concentration in surface waters would be *far higher*[1] than this: four parts per billion . . . Clearly the amounts of fenthion reaching the ground of target areas on Sanibel are *lethal, not merely dangerous, to crustaceans.*[1] Effects on the Refuge would vary depending upon the amount of fog received.

"Mr. Hicks reports that fogging removes not only mosquitoes but also wipes out the dragonflies, which are major predators of mosquitoes. Earlier studies made in Florida ponds reveal dragonfly larvae to be peculiarly sensitive to small amounts of fenthion, and this may mesh with Hicks' observation about the adults."

Stickle goes on to say, "What can occur has been shown in several of this Service's studies: small organisms affected by fenthion are eaten by birds which find them easy prey and the birds may be sickened or killed. This is secondary poisoning."

I'm very pleased to say that the response of both EPA under Mr. Train and the Department of the Interior under Mr. Reed was most gratifying and I am confident that this whole matter will be studied carefully. I was especially pleased at Mr. Reed's statement that he felt that the Department of Interior *could* exercise jurisdiction over drift onto USDI property. It is to be hoped that this whole matter will be resolved soon.

As we've already seen, fenthion-Baytex is highly toxic to birds. There was a case of mass mortality of migratory birds at Grand Forks, North Dakota, which was attributed to mosquito control operations applying the insecticide fenthion. The birds died because of the pesticide's toxicity, the method of application, and coincidence with the peak of the spring warbler migration. No less than 453 dead birds of 37 species were actually collected. But it was estimated that up to 25,000 birds actually died.

It is not surprising that such an occurrence took place, for fenthion is known to be used in bird control. All across the Sahel of sub-Saharan Africa, as well as southward into Kenya and Tanzania, large populations of Quela birds and their allies cause great economical loss in more than 20 countries. Fenthion as used in those countries is marketed as "Quelatox" and is actually used most successfully to kill birds.

Here we have a chemical, fenthion — call it Baytex, Quelatox, or what you will, it's all chemically the same. It is employed in Africa to *kill* birds and yet it is used on our Sanctuary Island, where birds are among our most important resources.

The EPA informs me that it is "unlawful to use any pesticide in any manner inconsistent with its labeling." I have in my posession a Baytex-fenthion label which states:

"1. This product is toxic to fish and wildlife. Birds feeding on treated areas may be killed. Do not apply when weather conditions favor drift from areas treated.

"2. Shrimp and crab may be killed at application rates recommended on this label. Do not apply where these are important resources.

"3. This product is highly toxic to bees exposed to direct treatment or residues on crops.

"4. Do not apply to ponds, lakes, and other bodies of water containing valuable fish.

"5. Warning: may be fatal if swallowed, inhaled or absorbed through the skin. Rapidly absorbed through the skin. Do not breathe sprayed mist".

All, I repeat *ALL* of the above label restrictions have been violated year after year on Sanibel Island. I have never seen a day when weather conditions did not favor a drift of poisonous spray from treated areas. Shrimp and crab ARE important resources throughout this region, in our bays and in the Gulf.

Bees, the source of Florida's multi-million-dollar honey industry, ARE killed in great numbers by this treatment. One apiarist in 1975 brought six hundred hives to Sanibel and got sixty barrels of honey. Some years before, this same man got eighty-five barrels of honey from only four hundred fifty hives on Sanibel. (The United Nations' Food and Agriculture Organization has declared the honey bee to be seriously endangered worldwide. Their concern is not so much honey production as pollination retardation — for bees make their greatest contribution to agriculture through pollination.) It is not uncommon to see literally hundreds of thousands of dead bees at the half dozen or so bee yards on Sanibel or at the edge of the water

---
[1]**Emphasis mine**

along the Gulf beach a day after a Baytex air raid.

Contrary to label restriction number four, application DOES take place over many freshwater ponds, lakes, and other bodies of water that contain numerous valuable fish.

And it IS inhaled by human beings who live here! Should visual disturbances, nausea, burning throat, or breathing difficulties result, it is not surprising because the organophosphates used today as pesticides represent fallout from the German nerve gas technology that was developed during World War II and acquired by the rest of the world soon afterward.

For a number of years I was closely associated with Bruce S. Ott, D.V.M., who was chief of the veterinary medical department at Edgewood Arsenal. Dr. Ott taught me a lot about this subject and the physiology of nerve gases and their close relatives, the organophosphates that are in such wide use today as pesticides. And *this* is what we are scattering about the world in which we live.

In the normal neural physiology, acetylcholine acts as the transmitter of the nerve impulse between the motor nerves and the skeletal muscles. Cholinesterases are enzymes that control this transmission. They can be blocked by anti-cholinesterase compounds, such as the many organophosphates. This large group of compounds, called OPs for short, all have anti-cholinesterase activity. That is, they inhibit the cholinesterases in the body. When cholinesterase is blocked by an anti-cholinesterase compound, then the transmission of the nerve impulse will continue via the acetylcholine and will fatigue and eventually paralyze the muscle. Eventual death of the organism can result (and in the case of insecticides and pesticides usually does). This process can be observed by the disoriented movement that is often visible in many creatures after exposure to organophosphate poisioning.

ABATE

The third organophosphate used here, Abate, is also used in violation of its label constraints. Abate is applied by helicopter as a larvicide. Its label shows many of the same prohibitions as does the Baytex label, and its use is equally illegal, for let us repeat the admonition of the EPA that it is "unlawful to use any pesticide in any manner inconsistent with the labeling."

The Abate label states:

> "This product is toxic to birds. Shrimp and crab may be killed at application rates recommended on this label. Do not apply where these are important resources. This product is toxic to bees and should not be applied when bees are actively visiting the area."

In July, 1972, Maurice Provost, entomologist of the Florida State Division of Health, wrote this about Abate:

"The Division of Health has recommended against Abate as a larvicide. But it has no jurisdiction over what a mosquito control district does. This unwarranted use of Abate by one district is the most flagrant nonconformance with State policy in two decades."

The district referred to was Lee County. Provost's concern related to possible genetic selection for resistance to Abate.

The Sanibel studies by both EPA and USDI are, of course, tremendously valuable and will no doubt prove to be of great interest. But the fact that EPA would ever allow the initial activities which are contrary to their own regulations to occur in the first place is in itself hard to understand and would indicate that EPA is not conscientiously performing its environmental protection function.

There is much concern for the human population here which is subjected to such frequent applications of toxic organophosphates. I discussed these matters with Morton S. Biskind, M.D., of Westport, Connecticut, who is a nationally-recognized authority on the use of pesticides and their effects on human beings. He was instrumental in getting the current national restrictions on DDT passed and also in fighting for pesticide laws in aerial pesticide application in New England. I quote his February 16, 1977, letter to me:

> "I am amazed to learn that anyone would use Fenthion in or near a wildlife refuge, since it is extremely toxic to birds, and under the name 'Quelatox' it is used specifically to eliminate 'undesirable' avian species. As for Abate, Pesticide Abstracts (EPA), August, 1976, has an abstract of a Hungarian study (76-1991) by Tevan and Lang (Effect of Abate mosquito larvicide on aquatic ecosystems) in which they used 1 per cent Abate ad-

sorbed on coarse sand. The abstract concludes, 'Even the relatively mild effect of Abate on aquatic ecosystems indicates that this larvicide should not be used in waters containing living organisms . . . '

"Fenthion would indeed be a hazard to humans and other mammals. It is at least 10 times as toxic as malathion, for instance, and 80 or 90 times as toxic for birds. It is one of the most persistent of the organophosphorous compounds, being 100 per cent effective against mosquitoes for as long as 42 weeks when applied to the side of a barn, and effective against insects that damage stored products for as long as 16 months (W. T. Thomson; Agricultural Chemicals. I. Insecticides, 1976, p 144-45). In a study in Spain on contamination of the environment with pesticides, Fenthion was the only organophosphorus compound detected. And dead birds were found during application of Fenthion to a rice field. (Pesticide Abstracts 74-2083).

"Japanese investigators have found that as little as 0.005 mg. Fenthion per kg. body weight (1/60,000th the lethal dose for 50 per cent of the animals) causes changes in the electroretinogram, with consequent disturbances in vision (Pesticide Abstracts 74-1420, 74-1979). And Fenthion is closely related chemically to Fenitrothion (Sumithion). The latter has been implicated by Canadian investigators in the production of Reye's Syndrome, a serious and often fatal encephalopathy of children. They demonstrated that Fenitrothion increases sus-

The author approaches a downed helicopter to check for Abate leakage into a fresh-water habitat. Fortunately the tanks were both empty and intact so no damage was done. A serious potential problem exists, however. If ever a Batex-laden DC-3 or a helicopter loaded with Abate crashes in a sensitive environment, we can expect serious damage. This has already been documented at Bimini, Bahamas, where a lot of Abate was dumped in a small area. That area remains a biological desert.

Photo, Mark Twombly, *Island Reporter*

ceptibility to an otherwise innocuous virus. (Crocker et al.: *Lancet* July 6, 1974.) Later these investigators also implicated the solvent. If anything Fenthion is more toxic than Fenitrothion.

"Thomson (ib.) says of Abate: 'Young shrimp or crabs in treated tidal waters may be injured or killed. Toxic to fish and birds. Toxic to bees. Do not use on pasture crops used for forage.'

"An island should provide an ideal environment for elimination of mosquitoes by release of sterile or incompatible male mosquitoes, which has been successful wherever tried."

In the early months of 1977, Reye's Syndrome was much in the news here in Florida. Although not a "reportable" disease, it was the subject of case data recently issued by the Communicable Disease Center, Atlanta, Georgia. This relatively new disease usually attacks children of eleven years or younger. It is often fatal or, short of death, leaves the victims little better than living vegetables. One researcher likened the present occurrence of Reye's Syndrome to polio long years ago, when it was first known as an obscure disease in Sweden. The obvious concern is that the incidence of Reye's Syndrome may explode as did that of polio.

I firmly intend to continue my opposition to the illegal and unreasonable use of deadly poisons on Sanibel Island. In 1974 I suggested a number of alternative methods of mosquito control, among which were the following:

Predators: We already have *Gambusia* and several other larva-eating fish species such as *Fundulus* spp., at least one of which lays eggs capable of survival in earth that is almost completely dried up. There are many others worthy of study. We have several species of dragonflies, but we should look further into the whole subject of predation. A study of salt-tolerant dragonflies should be made. Some species of planarians, which are soft-bodied flatworms, are being studied in an effort to use their predatory habits against mosquito larvae.

Parasites: There are some nematodes that offer promise in this field. (And they are *not* the nematodes that are spoiling your flower and vegetable beds.)

Pathogens: Opportunities for the use of protozoa, bacteria, fungi, and viruses that produce diseases in mosquitoes should be explored.

Attractants: Substances of use to attract mosquitoes to traps, to acceptable insecticides, to cultures of disease-producing agents, or to chemicals that will cause sterility is another possibility.

Sound Attraction: Another method, neglected here, concerns the use of the sound emitted by the female mosquito to attract the males to traps. This was developed many years ago at The Cornell University School of Medicine. It has been used with success in Cuba.

Induced Genetic Characteristics. Artificially raised mosquito populations carrying inherited characteristics unfavorable to survival might be released to breed with, and thus weaken or destroy, wild populations.

Hormones to Adversely Alter Metamorphosis. Insect growth regulators (IGRs) are now available from several sources. IGRs produce unfavorable changes or weaknesses in the development of mosquitoes and other pests. Since 1974 the Lee County Mosquito Control District has employed several highly-trained scientists who are pursuing a number of the above routes as well as other imaginative possibilities. It is my hope that successful methods will result that will satisfy all points of view: safety, protection of human and environmental health, and protection from too much abuse by our numerous mosquitoes.

At the same time we must all bear in mind the fact that was brought out very strongly by John Clark in his *Sanibel Report,* published by the Conservation Foundation of Washington, D.C.: that we must never underestimate the important niche that mosquito larvae of Sanibel Island occupy in the maintenance of the food web, and upon which the Island's very economic livelihood depends. People will not come to Sanibel Island if it is barren of wildlife. The birds feed upon the fish and the fish feed upon the mosquitoes. The Darling Wildlife Refuge will be a sorry place indeed when and if the day arrives when mosquito control is so successful that the war against the pesky insects will have been won.

That day is not yet here, for with all the lethal weapons employed, mosquito control as practiced on Sanibel Island today doesn't really work very well. At the height of the mosquito season at any time in recent years, it has not been comfortable to be out of doors. I suggest that it is not, from a comfort point of view, very much better to be parasitized by 60 mosquitoes, as was often the case during 1975 and '76, than it was in 1935 to be subjected to the attention of 600. The discomfort factor between 60 and 600 is hard to measure. Both are very uncomfortable.

Here on Sanibel we live in the midst of a crisis of conflicting values. Today we are attempting to slow rampant development and are succeeding to some extent. After all the chips are down, will some of the natural scene remain? Certainly some, but what remains would get a great boost if we were to get out of Lee County's environmentally brutal chemical mosquito control program in favor of some of the suggested alternatives. Any environmental manipulations should be designed to protect natural ecological processes — those dynamics that insure the stability and health of Sanibel Island, its wildlife *and* its human population.

12 *THE CATTLE EGRET* **Venturesome African Immigrant**

On Sanibel more than two hundred seventy bird species have been recorded at one time or another. On any one day during Spring or Fall migration perhaps 100 could be found without too much difficulty. The Christmas Bird Count of 1976 produced 119. This book on Sanibel wildlife is not going to emphasize birds because there are many bird authorities on the Island who are much better qualified than I to discuss this subject. Also, much has already been written on our birds. So I will mention only a few of my favorites.

One bird that really turns me on is the Cattle Egret. This cosmopolitan creature has many names: in South Africa it is called the Tick Bird; in India the Surkhia Bagla; in West Pakistan, the Gai-Bagle; and in France, Heron Gard-Boeufs. Its technical name is *Bubulcus ibis,* and what a marvelously competent creature it is!

In these days when the world is concerned with endangered species, when so many creatures from the Polar Bear and the Blue Whale to the Tiger and the Cheetah are threatened with extinction, when we here on Sanibel are worried about losing many of our own species, the Cattle Egret, in common with a few other animals, thrives in company with man, his commensals and his civilization.

The Cattle Egret is native to Africa where it

The Cattle Egret in his original environment.

is found throughout much of the sub-Sahara region all the way to the Cape. It is also known to be an old resident of Madagascar. In the last century, however, the Cattle Egret took matters into his own "hands" and things began to happen. It emigrated to Asia where it is very common today on the Indian sub-continent and in Malaysia and IndoChina. It is also known in Australia, New Guinea and other areas of the Australasian region.

It is believed that in the last century the Cattle Egret also left its native Africa for regions to the north, for today it is seen in Spain and Portugal and has been observed breeding in France. It has been found in Britain, Denmark, Hungary and the Balkans.

Perhaps one of the most remarkable ornithological events of recent times took place about the turn of the century when somehow the Cattle Egret crossed the Atlantic and landed on mainland South America. It is believed that the point of landing was what was then the colony of British Guiana, now known as Guyana, on the northeastern coast of the continent. From there, in the last seventy-five years or so, the Cattle Egret has dispersed deep into South American regions as far south as southern Brazil and parts of Argentina and west to the Pacific Ocean, including the Galapagos Islands.

During this same period the species moved to most of the West Indian Islands and the Bahamas where it is very common in many places, and on to mainland North America where it may be seen in most of the eastern seaboard states and at least as far into the interior as the Great Lakes states. It can also be found in the Maritime Provinces of Canada and in Ontario.

Wherever this remarkable animal moves into new areas of the world, it joins up with either domestic grazing stock or wild herds of large animals. Contrary to the impression that might be deduced from the South African name, this is not a bird that removes ectoparasites from ungulates but, in fact, it feeds mostly on the ground on insects, lizards, rodents, and other small creatures which are disturbed by the moving cattle or herds of wild animals. In India and in Trinidad it may be seen riding on the backs of Water Buffalo; in Wisconsin, among the dairy herds. In Africa you can see it riding on, or feeding among, Rhinoceros, Elephant,

Cape Buffalo and the many other hoofed animals of the veldt.

On Sanibel, where we have few hoofed animals, the Cattle Egret hunts lizards and insects on vacant lots and filled land. It has, however, discovered a satisfactory substitute for herds of plains game — it follows mowing machines that cut the grass on the shoulders of our roads and highways, exposing grasshoppers and other small prey.

The Cattle Egret is seldom seen among the other egrets and herons of the "Ding" Darling Refuge. Following its discovery in the United States, some ornithologists believed that it would adapt to a niche that did not compete with its related colonial herons and egrets. However, LeBuff has seen this bird seriously compete with its relatives at such heronries as Upper Bird Key and others in this region.

Africans recognize the value of this bird as an insect-control agent and in some areas it is not disturbed or killed. Ornithologists tell us that the Cattle Egret is one of the most primitive of the Ciconiiformes, the group that includes the herons and their allies. It is interesting to note that, although primitive, the animal is much more successful than many of its more specialized relatives. Consideration of the guiding principles of evolution leads one to understand why: versatility and adaptability are qualities that lead to success.

Many bird-oriented people on Sanibel tend to deprecate the Cattle Egret. It's not really a very much loved bird. Some people discuss it in the same sentence with European Starlings and House Sparrows. But for my part I think it's a very remarkable and fine bird. Its amazing recent history can only be admired. The Cattle Egret may well be the only one of the herons left here in a few years if we continue the destruction of our Island's environment. Let's all give this versatile bird the place in our consideration that it so richly merits.

Wood Stork. This frequent Sanibel visitor is often erroneously called "Wood Ibis" but is, in fact, the only true stork native to the United States.

13 *SHARKS AND RAYS*
Or: It's Safer to Swim in Your Pool

Not so long ago the Lee County commissioners outlawed shark fishing from county beaches, and one commissioner was quoted as saying that there really weren't any sharks around here anyway. Such an attitude is ridiculous, as every fisherman or casual observer knows. Ignoring the fact that there are sharks in our Gulf waters does as much injustice to the truth as does the irrational fear of sharks that is so widespread.

Millions of people swim and bathe in the waters of the Atlantic and the Gulf of Mexico every year, and recorded shark attacks are rare. But make no mistake, shark attacks do occur in Florida — there were four of a serious nature in 1976. Also, two shark attacks have been recorded over the years at Sanibel. Those great fish are indeed out there and they provide a fascinating subject for study for the naturalist and sport for some avid fishermen.

Dr. Gordon Hubbell, Director of the Crandon Park Zoo at Key Biscayne, Florida, is a dedicated shark fisherman and one of his favorite locations for indulging his pastime is Sanibel's famed beach. Here he has taken Bull Sharks, Lemon Sharks, Blacktip, Sharp-nose, Nurse, Bonnethead and Hammerhead Sharks. He is still trying to get a Tiger Shark, but this great creature, the "garbage can" of the sea, has so far eluded him, although it is known to occur in these waters.

One of the most interesting species is *Carcharhinus leucas,* the great Bull Shark. This species is particularly fascinating because it lives in both marine and fresh water habitats. It is found in the fresh waters of Lake Nicaragua in Central America and also travels far upstream, as much as one hundred fifty miles, in many African rivers.

My son, David, and I recently looked into the shark fishery in Lake Nicaragua and found sharks to be very plentiful there and also to be greatly feared by the local citizens. There are many authentic records of shark attack in lake Nicaragua. The Bull Shark may attain a length of ten feet and a weight of five hundred pounds. This is the same species that we have near Sanibel's beach, although a few ichthyologists assign separate species status to the freshwater animal. The saltwater population, which ranges from the Carolinas to Brazil on the Atlantic coast, and throughout much of the Pacific as well, is not noted for a large number of recorded shark attacks.

One of the largest sharks that Dr. Hubbell has ever taken is a three hundred eight-pound Bull Shark, caught when he was fishing from the beach at Sanibel Lighthouse.

Another shark that is common in our waters is the Lemon Shark, *Negaprion brevirostris.* This species is commonly found inshore in shallower water. It will also sometimes enter fresh waters such as upstream in the Caloosahatchee and is known to attack humans.

The Lemon Shark is yellowish in color; the upper part may be brownish to yellow, and the under surface, as in most sharks, is white to beige. It has two dorsal fins, the second one being almost as large as the anterior one. Some Lemon Sharks attain a size of about eleven feet and three hundred pounds. Dr. Hubbell caught a seven-foot-eleven-inch Lemon Shark weighing one hundred eighty-five pounds in Sanibel waters. Two others caught here weighed one hundred forty pounds and one hundred seventy pounds.

The Blacktip Shark, *Carcharhinus limbatus,* and its allied species *C. maculipennis,* are closely related and often confused. Both are here and both are very good to eat. The former is the Lesser Blacktip and the latter is the great Atlantic Blacktip, sometimes called the Spinner Shark because of its wild rotation

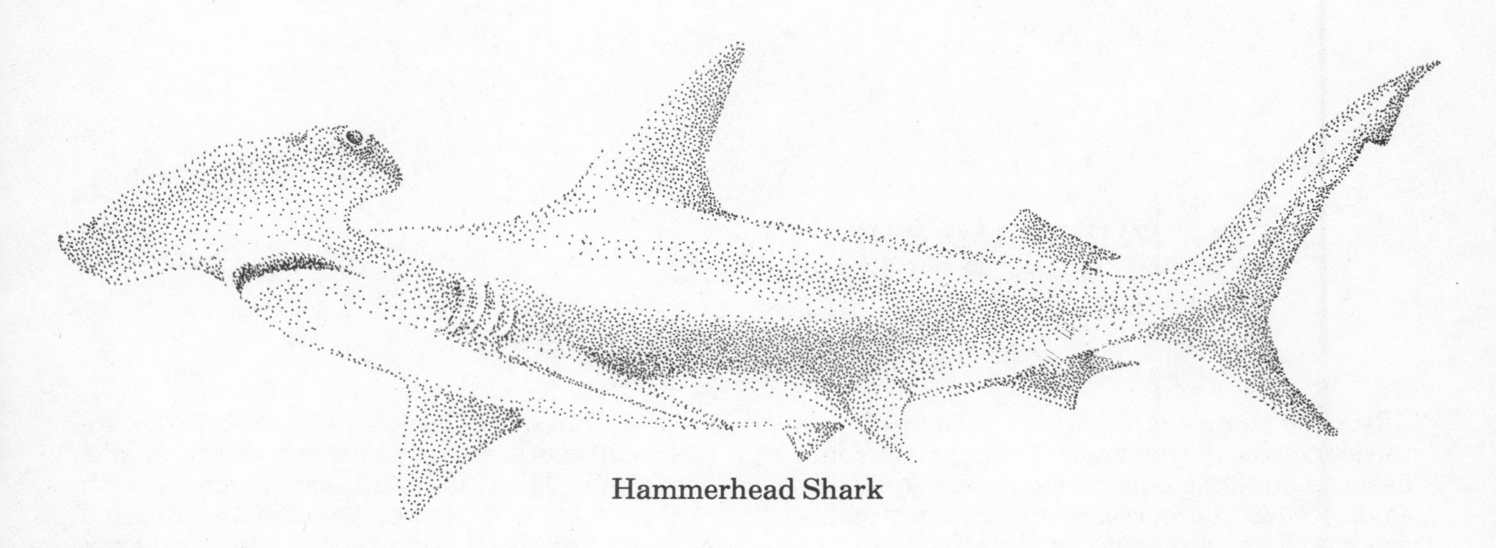
Hammerhead Shark

when hooked.

The smaller Blacktip attains only about seven feet. A record animal, caught just north of Sanibel at Boca Grande in 1973, was six feet eight inches. The Spinner gets substantially larger and is thought by many to be the best-tasting shark in our waters.

Another interesting shark group is the family Sphrynidae, the Hammerheads, of which there are several species. The Hammerheads are aggressive fishes, known to attack man. A large Hammerhead specimen is often seen near Boca Grande Pass. Local fishermen tell legendary stories about this huge individual, which may in fact be several different sharks living in the same general area.

Hammerheads are readily identified because the head extends laterally in the form of a flattened, somewhat thickened, "hammer" at the tips of which the eyes are located. In a huge specimen, the eyes may be three feet apart. Hammerheads attain a length of almost eighteen feet and may weigh up to three quarters of a ton.

The writer once saw a remarkable sight while boating in Florida waters. A heavy sea was rolling off Salerno Beach on the East Coast. A great fifteen foot wave rolled toward the boat and offered its broadside surface to view like the window of a great aquarium. At the instant before the wave broke, a huge Hammerhead flashed through the wave at a height some six feet above eye level. It is seldom that one can view a great fish swimming in such a way.

Salerno Beach used to be the site of a very successful shark fishery some forty years ago when there was a market for shark leather, for dried shark fins for export to mainland China, and for shark liver from which was extracted vitamin-laden oil. I used to visit this fishery and go out as crew on the shark vessels. Half-mile-long chain trotlines were used. These were anchored and buoyed at each end. Every fifty feet there was a long chain leader with a huge hook. We would pick up the first buoy and the first anchor and run it across two large pulleys on the gunwale of the fishing vessel, set the engine at a slow speed, and the boat would then steer itself along the chain line.

Every few minutes a leader would surface, usually with a huge shark hooked. The sharks were gaffed and winched into the hold. After a day's activity many tons of sharks would fill the hold.

Sometimes a huge shark would appear with only the head remaining, the rest having been chomped off by an even larger shark. Sometimes a small shark would be hooked and before we got there it would be swallowed by a huge monster shark, so that we would get two sharks on one hook.

At dockside after a long day of fishing, a crew of specially imported and highly-skilled West Indian shark-skinners were on hand. With great dexterity they would skin all of these tons of sharks in a short time. The fins were salted in barrels and the livers dressed out and put aside for later rendering. The remainder of the animal was ground up and dried for fertilizer. This industry died with the rise of Red China and the concomitant loss of the Asian fin market and, more importantly, with the advent of synthetic Vitamin D in the pharmaceutical industry.

There are two groups of immense filter-feeding sharks that are quite gentle and feed only on plankton. These are the Basking Sharks and the Whale Sharks. One huge Basking Shark, *Cetorhinus maximus,* once produced a liver that weighed eight hundred sixty pounds. There is a huge quantity of Vitamin D in a liver that weighs eight hundred sixty pounds! This animal reaches a size of over thirty feet.

The Whale Shark, *Rhincodon typus,* attains the unbelievable length of forty-five feet. One weighed in at more than twenty-six thousand pounds! It is, however, quite a docile animal. A friend in Panama regularly swims among these sharks, studying their habits.

I recently talked to a leading shark authority who told me that sharks in murky water are more prone to bite the extremities of a victim than in clear water. He stated that Cuban waters are generally murky and there are many records of shark attacks in the waters of that island. In the Bahamas where the waters are crystal clear there are fewer records of shark attacks.

In murky nutrient-laden waters, such as those of Sanibel, one should be at least cautious and on the lookout for shadows and fins. Shallow water is not a deterrent to a hungry shark.

Dr. Hubbell's shark fishing technique may be of interest to some. He uses fresh bait—jack or fresh caught bloody Ladyfish. He emphasizes that dead bait, such as an old fish head, is of little use. He catches most of his sharks on light tackle.

A large Lemon Shark was caught in June of 1976 at Sanibel by Thomas Carr, using Catfish for bait when fishing in San Carlos Bay with a large hook and heavy nylon line. The shark was eight feet, 10 inches long and is pictured here.

Alice Kyllo found a large pale-colored Hammerhead Shark at Bowman's Beach in 1976. It received a lot of attention because some thought it to be an albino, but I personally believe it was a normal Hammerhead that had bleached out in the sun.

The Great White Shark, *Carcharodon carcharias,* about which so much exciting fiction has been written recently, is unlikely ever to be found in Sanibel waters because of its pelagic nature. They are in the Gulf, however, and several have been taken into Mote Marine Laboratory by scientists of that institution. This species' record was one 36½ feet long. A 21 foot animal weighed 7,100 pounds. Its liver was 1000 pounds in weight.

Another great group of related fishes are the rays and skates, and they are well represented in Sanibel waters.

Sting Rays are among the least popular members of Sanibel's wildlife community and for a very good reason to which I can personally attest. Some years ago, before we had any medical services on this Island, I managed, with great skill and dexterity, to step down hard on a large Sting Ray at the Rocks beach. I immediately felt an absolutely excruciating pain rather like you would imagine a red hot poker jammed into your foot would feel. This pain lasted for many hours but was about 90% gone by the next day, after which a small painful sensation lingered on for quite a few months.

Starting in April each year there are many Sting Ray accidents on Sanibel. One clinic treated fifteen one May weekend. They seem to taper off by mid-summer and are reduced to just a few isolated cases. All of this means that the ray and its allies are well-known, if rather unpopular, members of Sanibel's wildlife list.

Perhaps the most common group of rays that can do you damage are those which belong to the family Dasyatidae and of these, three are known in these parts and they are all members of the genus *Dasyatis* — the Southern Sting Ray, the Stingaree and Say's Sting Ray. These dangerous animals, together with all other rays, skates and sharks are Elasmobranchs. This designation denotes the primitive group of fishes that do not have bones but have cartilaginous skeletons. The skates and rays belong to an order called Batoidei and there are all together some three hundred fifty species in all the seas of the world, and also quite a number that enter fresh water river systems of South America, Africa, Asia and even North America.

The ray's anatomy is interesting in that the pectoral fins are attached to the sides of the head with the gills underneath them. These great pectoral fins then are what we refer to as the "wings" of a ray. With this anatomical development these beasts are able literally to "fly" through the water. On sandy or muddy bottoms, they bury themselves by fluttering the wings and allowing the sedimentary material to rest on top. In clear water it is

Photo, *Island Reporter*

Thomas Carr and his eight-foot-ten-inch Lemon Shark

possible to see the outline of the ray on the bottom. But in waters such as we have on both sides of Sanibel, this is not often possible, making it mandatory or at least very good sense to shuffle your feet in order to avoid receiving a painful wound from one of these animals.

The Sting Ray tail is usually quite flexible and much longer than the body. On the dorsal surface of the tail can be found the venomous spines — commonly just one, but sometimes as many as three. Some students feel that there is a constant replacement of spines much as the rattlesnake fang is replaced. This makes for the normal complement of one, but if the new one takes the place of one not yet shed, that makes two and sometimes this process leads to more.

This flexible appendage can be whipped about in most directions and is readily lodged in your leg if you're careless enough to step on or get too near such an animal. A large ray can drive his spine right through the planking of a small boat and there are cases of a person's leg being completely penetrated, in one side and out the other, by a large Sting Ray spine. Most accidents occur to the foot, ankle or calf but there are records of abdominal penetration and such rare cases have always been fatal.

I have searched the literature and have talked to several doctors skilled in treating the stings of rays and have found the following to be an acceptable treatment:

1. Irrigate the wound with a solution of Epsom Salts.
2. Surgically remove any of the venom spines, together with any foreign material that might remain in the wound.
3. Immerse the wound in extremely hot water — as hot as one can stand — for upwards of an hour. The theory here is that the hot water oxidizes, or at least causes the chemical break-down of, the toxicant.
4. Make a second search of the wound for further bits of foreign tissue.
5. Suture if required. Some doctors also give an anti-tetanus injection.[1]

Forty years ago, before this writer realized that conservation was going to become an important consideration for the future of the human species on this abused planet, I participated along with other Floridians in an illegal and environmentally destructive activity. It was in the days when the "runboat" used to come down from Charlotte Harbor, stopping at numerous fish houses on the way down to Punta Rassa, picking up fish and leaving ice.

The many lone fishermen who spent their days angling for Redfish and Trout among the islands and coves of the Bay were the backbone of the fishing industry. But there were some who were impatient with this slow tedious way of making a living and wanted to catch more fish with less effort. I was for a short time associated with such a group. We undertook an illegal activity known as "stop-netting." The procedure was to put a half-mile-long tarred net across the mouth of a river or bayou at high tide and wait for the out-rushing tide to enmesh all the animals of that enclosed area. Thus at low tide, thousands of fish would be entrapped.

Upwards of a dozen men would then get out onto the flats and by hand haul this great net into a small bunt perhaps 25 feet in diameter in which there would be literally tons of fish of all kinds: Shark, Mullet, Tarpon, Redfish, Trout, Snook, Catfish and hundreds and hundreds of Sting Rays and Stingarees. The procedure then was to stand in the waist-deep water beside this last bunt enclosure and try to release the trash fish while saving the Redfish, Snook and Trout. Among the most abundant of the trash fish were the Sting Rays. They were released by wrapping the very tip of the Ray's tail around one's finger and, using this handle, gently easing the creature over the top of the cork line and releasing it at one's feet.

After an hour of this, one would be waist-deep in Sting Rays, a very uncomfortable thought, but really not so dangerous because by the time they had been pulled up in this huge net and had been traumatized by the manhandling, they were rather docile and thoroughly fatigued and in no mood to go around stabbing their tormentors. As I recall, no one was stung by a Ray although we handled many hundreds in this way.

The practice of stop-netting is no longer tolerated here in Southwest Florida. I suppose I should feel ashamed at having been a part of such a destructive business, but as a youngster in high school I didn't really know any better. I

[1]There is an old Florida folk medicine procedure for the treatment of Sting Ray wounds: Soak a clean, boiled cloth in strained honey and wrap the wounded limb in this saturated cloth. Change daily for three or four days.

guess that is not really true either because I remember on more than one occasion we had to douse our lights and crouch low in the water when we feared the "prohis"[1] were passing near.

In some cultures the dried spine of the Sting Ray is used as a spear or arrow point; in others, the Sting Ray pectoral fin or wing is eaten. One can order Sting Ray in restaurants in Southeast Asia.

A considerable quantity of Ray meat is also eaten in the United States, but most people are not aware of it. Many of the "scallops" sold in this country are not scallops at all, but Sting Ray fins that are cut into round discs by using something like a cookie cutter. This is a very widespread practice. It smells of misrepresentation to me and I am surprised that the FDA has not looked into this.

---

[1]**Florida slang of that time for "the law." The term derived from "prohibition enforcement" officers of an earlier day.**

There are many other species of Rays which we won't discuss here. But there are a few groups which we must touch upon, such as the Sawfish which is equipped with a great tooth-armed beak protruding from the front of the head. This Ray is found in many tropical regions including this one and it can do a lot of damage with its saw-bill as it thrashes it from side to side.

Another group includes the Devil Fish or Manta Ray which is really a docile creature in spite of its huge size. Manta Rays are not flesh eaters, but feed on plankton, the small, often microscopic, plants and animals that live floating in the seas. Like the Whale Shark and Basking Shark, it is equipped with a device to strain huge quantities of sea water and thus secure sufficient amounts of plankton to provide for its nutrition. Mantas can grow to be twenty-two feet across the wing tips and reach a weight of more than thirty-five hundred pounds. They are great jumpers and it is a

A close call with a Southern Sting Ray

truly horrendous feeling to be in a small boat next to one of these creatures when it comes crashing down beside you. In days past, they were very evident along the shores of Sanibel but not many are seen now. They can be spotted easily as they swim lazily along the surface, flopping the tips of their wings out of the water.

One of the most fascinating of all the Ray groups is that of the Electric or Torpedo Rays, of the family Torpedinidae. A special issue of Time Magazine dated July 4, 1776, carried an interesting story on the Torpedo Ray on page 63. This article describes how a group of investigators linked together electrically "felt a commotion" as the Ray discharged his electricity. Dr. John Hunter's experiments are discussed. He dissected the Electric Ray in that year, 1776, and discovered the electricity-producing organ. Dr. Hunter was physician to King George III.

There are over thirty species of these remarkable animals which are equipped with paired electrical organs consisting of modified muscle tissue. The organs are located in the wings and are made up of discs of plates of which there are many hundreds in each organ. The top of the Ray is electrically positive and the bottom negative. A Torpedo Ray is capable of discharging a wallop of more than 200 volts. Electrical discharge is used to overcome active prey food, for orientation much as sonar is used, and possibly as a defensive device. Electric Rays have been hooked up to electronic devices to control radios or lighting fixtures. Such a demonstration can be quite dramatic. I have myself rigged a loud speaker for this purpose. You simply scrape electrodes on the animal to make the speaker growl.

These animals are not nearly so common in our waters as are the Sting Rays, but they are present and of considerable interest.

The Sting Rays and the Electric Rays, like the Manta Rays, all give birth to living young. That is, they are ovoviviparous. The egg cases found on our beaches and commonly called "Sea Purses" or "Mermaids' Purses" belong to the Skates, another group that is closely allied to the Rays.

14 *SNAKES ALIVE*
The Ophidians of Sanibel Island

Fifteen different kinds of snakes are known on Sanibel. Unlike the lizards and frogs, of which there are exotic introductions, these are all native Florida snakes.

In 1975 LeBuff and I attempted to evaluate population trends of many Sanibel animal species. Our conclusions in regard to the snakes were: one kind seems to be increasing in number; five are declining and nine are holding their own despite over-development and heavy automobile traffic.

There are two kinds of local water snakes — the Florida Water Snake, which lives in the fresh water system of the Sanibel River, and the real estate lakes; and the Mangrove Water Snake or so-called Flat-Tailed Water Snake.

They are both the same species, *Natrix fasciata,* but the fresh water form is called *pictiventris,* which describes its rather delicate and interesting belly pattern. The so-called Flat-Tailed, or Mangrove, Water Snake carries for its sub-specific name the word *compressicauda,* which translates into "flat-tailed."

I have examined literally hundreds of specimens of this snake and have yet to see one with a flat tail. I've perused the literature and asked questions of fellow herpetologists and the best conclusion I can come to is that the type specimen (the original specimen used to describe this subspecies) had been run over on the road and had its tail flattened by the offending vehicle. Zoologists make strange judgments like that sometimes, but due to the priority system of taxonomy, the misleading name, *compressicauda,* will probably carry on forever.

The Mangrove Water Snake can be seen out on the J. N. "Ding" Darling National Wildlife Refuge, often on the sides of the dike road, sometimes in among the mangrove roots. It occurs in two different color phases — a red one and a black, or gray one. Sometimes there are intergrades. The reddish one is quite beautiful — a brick-red color. It is a keel-scaled (rough skinned) snake, with round pupils. It feeds principally on small fish and produces living young.

I had a large female Mangrove Water Snake in 1975 which, on June 25, gave birth to eleven living young, about half of which were dark phase and half red. I retained two babies in my laboratory, a red one and a dark phased one. Today they are young adults that savagely attack the *Gambusia* and other fish placed in their aquarium. Sometimes one of the snakes will grasp and swallow a fish from the head end while the other starts to swallow from the fish's tail end. An exciting confrontation ensues. "Red", as I call him, is a bit larger than "Blackie" and is able to win the fight by swallowing not only the fish but a third of Blackie as well. Blackie then releases his end of the fish, thrashes about a bit and pulls himself out of Red's gullet.

Charles LeBuff, who has studied the herpetology of this Island for many years, informs me that the Mangrove Water Snake used to be found in considerable abundance; today one must search to find a specimen.

A third water snake, the Florida Green Water Snake, was recorded here only once, back in 1962. In the days before herbicides and hyacinth weevils, great rafts of water hyacinths used to drift down the Caloosahatchee, carrying with them many fresh water animals that sometimes became stranded on barrier islands. It is probably by this route that the 1962 specimen found its way to Sanibel Island.

The Florida Brown Snake was once very plentiful here. This tiny creature, about as big around as a pencil, brown in color with dark spots, used to be visible on the highways in the summertime after a rain; but I have found only one specimen during the past two years. The young are live-born during the summer and become food for Coral Snakes, Indigo Snakes, Black Racers, Coachwhips and Herons, as well as Hawks. The Florida Brown Snake feeds on small worms and other invertebrates.

There are two Garter Snakes included in our herpetofauna. One of them, the Southern Ribbon Snake, is our most abundant ophidian. This striped, very thin creature sometimes reaches three feet in length. It is seen in great abundance in the summer time, but unfortunately all too often as many as fifteen or sixteen will be seen smashed on a five-mile stretch of road on a hot rainy night here on Sanibel.

This live-bearing animal is preyed upon by raccoons, other snakes such as the Coachwhip and the Indigo, as well as by the Kestrel. The Ribbon Snake feeds on tadpoles, small frogs and minnows. Paul Zajicek, a knowledgeable young herpetologist of Sanibel, observed eleven of them one July night in 1975 after the first real rain of the season.

The other Garter Snake is the Eastern or "Common" Garter Snake, which many of us know from the north. This is a very rare animal and only one was seen here until 1974. In 1961 Charles LeBuff collected the first one ever to be identified on Sanibel Island. In October, 1974, I found a road kill on Sanibel which both LeBuff and I identified as this species. I sent it to the British Museum of Natural History with the 1976 "Sanibel Expeditionary Force" of the British Royal Army (a special army group that visited here from the United Kingdom) for further confirmation of our identification. On a visit to London in July, 1976, I learned that our I.D. was supported.

A beautiful tiny glossy black snake with an orange belly and a yellow ring around its neck is called, not surprisingly, the Southern Ringneck Snake, or Corkscrew. It is relatively abundant on the mainland but rather scarce here; one sees only six or eight specimens each year. Its diet consists of small lizards and small frogs, and since it is found in the main ridge areas where the Greenhouse Frog occurs, that creature is undoubtedly preyed upon by the Ringneck Snake. It, in turn, is fed upon by the Coral Snake.

In July 1975, Richard Workman observed an adult Ringneck Snake followed in the same track by a juvenile; both were crossing Sanibel Gardens road en route to the Post Office.

Three "Racer" snakes exist on Sanibel. The Southern Black Racer, which grows to more than six feet, is a sort of frosty-black color. This is a speedy animal, often heard rattling through the dry underbrush.

The Eastern Coachwhip Snake is a large creature which sometimes grows to be eight and a half feet long, but averages about five feet. It is tan toward the rear and dark toward the head. It is a very fast fellow and can be quite savage when protecting itself. If you should ever grasp one, expect to be bitten savagely.

The third "Racer", the Great Eastern Indigo, is North America's largest serpent. It grows to eight feet, seven and a half inches long, and perhaps even larger in some parts of its range. This is the only legally protected snake in the State of Florida. It is considered to be valuable in that it consumes other snakes, including the Eastern Diamondback Rattlesnake. But it also eats frogs, fish and small mammals. It is, in turn, eaten by raptors, raccoons and possibly possums. This impressive serpent is heavy-bodied, glossy blue-black, with reddish chin areas, and usually very gentle. It is almost unknown for one of these animals to bite — even if you were to seize one in the wild, it probably wouldn't bite you. For this reason there was a considerable commercial traffic in Indigos when they were shipped to fanciers in the north who liked keeping them because of their large size and gentle nature.

An interesting fact about this species is that it has a geographic range larger than almost any other snake species in the world. It ranges from here around the Gulf and into our Southwest, thence southward almost to Patagonia.

The Yellow Rat Snake occurs on Sanibel, its center of distribution being thought to be around the Old Bailey Store, now the Childrens' Center. But the only ones I've had any experience with were found way out on the "Ding" Darling Refuge near the bird tower. This is far away from any real fresh water and certainly in what has to be considered a salt water habitat. Yet elsewhere in its Florida range, it is an animal found in agricultural and other areas near fresh water. Normally the Yellow Rat Snake doesn't grow much over fifty inches, but the record is eighty-four. It is yellow with four dark longitudinal stripes.

In June 1973, the Fox, Haff and Culpepper families in "Sanibel Isles" watched a fas-

Noted wildlife photographer Laura Riley shooting a seven-foot-three-inch Indigo held by the author. The Indigo is the only snake protected by law in Florida.

cinating performance by a three-footer as he swam a salt water canal and shinnied up on an oyster bed. When his audience refused to go away, the snake apparently decided to go on about its business, which consisted of investigating the oyster bed, climbing the sea wall, crossing a graveled area and climbing a gumbo limbo tree, where it reclined on one of the branches in defiance of the onlookers who insisted on intruding on its privacy.

A close relative of the Yellow Rat Snake, the Corn Snake or Red Rat Snake, was recently positively confirmed to occur on Sanibel when I took a specimen at Tarpon Bay Road and Gulf Drive. One of Florida's most beautiful serpents, with reds, yellows, black, white and even pink sometimes, it is an animal that has adapted to human development. It, like the foregoing species, is a true constrictor. That is, it wraps around its prey and kills by squeezing to prevent breathing and circulation. Constrictors do not "crush" their prey, as the popular concept would have us believe.

Both the Red and Yellow Rat Snakes are useful to have around the garden, as they keep the large House Mouse population under control. Also, Roof Rats are a pest species here, as is the Norway Rat, but to a lesser extent. Both are consumed by our Rat Snakes.

Finally we come to the last three species of snakes on Sanibel Island, two of which are dangerous forms: the Eastern Coral Snake and the great Eastern Diamondback Rattlesnake. Both are deadly poisonous and both command great respect.

The Eastern Coral Snake is really not uncommon on Sanibel. One young Sanibel student of my acquaintance, Bob Watterson, is quite skilled at ferreting them out and has observed quite a number of them on the Indian Mound Trail of the "Ding" Darling Refuge, which, incidentally represents a rather typical Coral Snake habitat.

The Eastern Coral Snake is brilliantly colored with red, black and yellow rings that pass completely around the body. The color sequence is black nose, yellow headband, black head posteriorly and black neck, narrow yellow, broad red flecked with black, narrow yellow — and so on down the body to the tail which is alternate black and yellow.

Many people are bored with trying to remember the red, yellow and black sequence in each of the three Florida species of highly colored snakes and thus often confuse the two mimics with the truly dangerous venomous Coral Snake. ("Red next to yellow can kill a fellow; red next to black, I'm all right, Jack.") This problem can be easily and safely solved here in the Southeast, but only here, by simply remembering that of the three, only one has a black nose, and that is the Coral Snake.

The mimic forms, the Scarlet Kingsnake and the Scarlet Snake, have red noses. The latter has never been found on Sanibel, but a specimen of the former was taken by George Weymouth the evening of May 23, 1977, in the house of Sanibel's Mayor Goss. This is the first confirmed record of this species from Sanibel Island.

There are two major groups of highly venomous serpents: the Viperine snakes, including our Rattlesnake, and the Elapine snakes such as the Coral Snake. The Viper and Pit Viper have, in the upper jaw, two long hollow hypodermic needle-like fangs that fold back against the roof of the mouth. The Elapine snakes have firmly fixed hollow fangs in the upper jaw.

The greatest of all the Elapids is the great King Cobra of Asia, the largest and perhaps most aggressive and dangerous venomous snake in the world. But our little Coral Snake, being a member of this family, Elapidae, merits our respect and extreme care always when it is encountered in the wild. Contrary to many stories, it is not always docile and inoffensive, and accidents do occur.

The Scarlet Kingsnake captured by George Weymouth in the home of Sanibel's Mayor Porter Goss. This specimen was a "first" for Sanibel Island. It later escaped. The characteristic red nose doesn't show here.

Photos, Dallas Kinney, *Island Reporter*

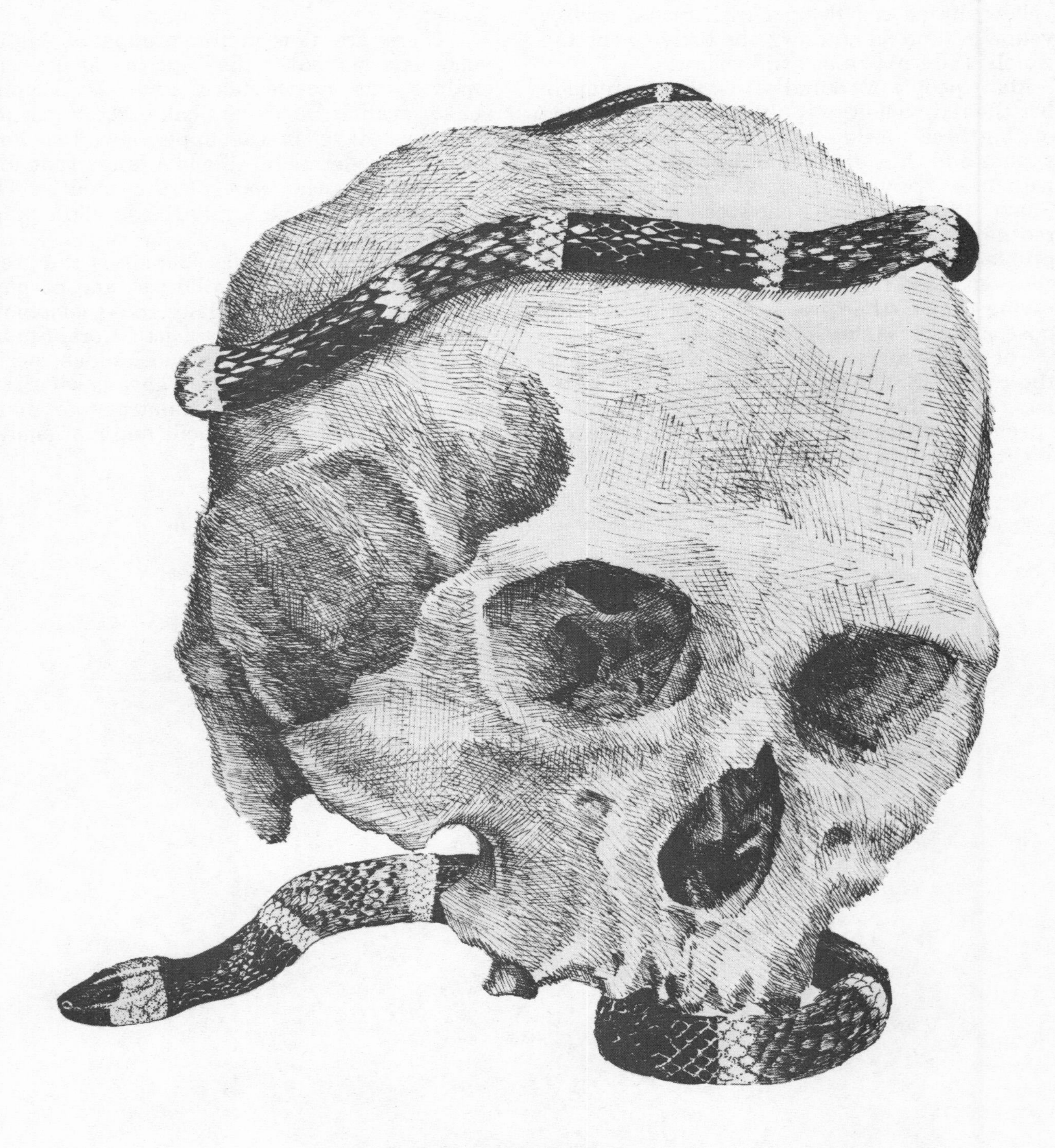

Eastern Coral Snake

Dr. Archie Carr of the University of Florida, had an astonishing and unpleasant experience with a Coral Snake. He states that he was walking through a hammock when a large Coral Snake struck him savagely on the leg. He jumped back and the creature lashed its body back and forth in quick lateral, jerky motions and immediately disappeared beneath the ground cover. This kind of behavior is not at all common, for the Coral Snake is normally shy and retiring.

Jim Vanas, formerly of the Everglades Wonder Gardens at Bonita Springs, used to show Coral Snakes, and the way he handled them made them seem to be about as dangerous as a string of beads. But I repeat, caution should always be used when one of these animals is encountered.

The venom of the Coral Snake is particularly virulent, being a powerful neurotoxin. For a long time there was no specific Coral Snake antivenin or serum produced in this country. Elapid antivenin produced at Instituto Butantã, São Paulo, Brazil, for the treatment of South American Coral Snake bites, was used in this country with success. Now Wyeth of Philadelphia makes a species specific product for *Micrurus fulvius,* which is the Latin name for our form. Sanibel's Island Apothecary keeps a supply of this antivenin (as well as that for the Rattlesnake) and now that the Island has good medical services, there is no reason for any fatal accidents to occur.

When I was a schoolboy in Florida, I took the world's record Coral Snake. It was fifty-two inches long, the absolute giant of its kind, three and one-half inches longer than the official record shown by the New York Zoological Society which keeps such records. Following quite an exciting history, my specimen rests today at Yale's Peabody Museum. The snake was turned up by a plow in the western part of Palm Beach County. It was quite docile and was easily put into a crocus sack (which is what gunny sacks are called in rural Florida). I brought it home for the amusement and pleasure of my family and neighbors.

Before I had an opportunity properly to preserve this dangerous animal for the Museum, it escaped in the house — unfortunately, on a day when my mother was having a bridge luncheon. I thought she handled a difficult and potentially explosive situation rather well, for when she went to replenish the logs that were burning in the fireplace on that cold day, she found the snake comfortably coiled up beneath a log on the hearth. She very gingerly replaced the log and let the snake rest in peace until the bridge ladies left. I was then summoned. The snake was recaptured and soon found itself in the pickle jar en route to the Museum. I still bear the scars from that traumatic experience, not from the snake, but from my mother!

The accompanying drawing was done from life with considerable courage by Molly Eckler Brown, using the second largest specimen I have ever seen — a forty-two-inch animal that recently appeared here.

The Diamondback Rattlesnake has two relatives on the nearby mainland — the Eastern Cottonmouth or "Water Moccasin" and the Dusky Pigmy Rattlesnake, all pit vipers. Only the Diamondback occurs on Sanibel.

Undoubtedly one of the most noteworthy items on Florida's wildlife "menu" has to be the Eastern Diamondback Rattlesnake, the largest of all Rattlesnakes and number seven on the list of the world's most lethal serpents.

Photo, Dallas Kinney, *Island Reporter*

This small Eastern Diamondback Rattlesnake washed ashore among bathers on Sanibel's Gulf Beach in March, 1977.

This is also one of the most beautiful Rattlesnake species, as may be seen from the accompanying drawing, made from a magnificent five-foot specimen.

Clarence Rutland, a long-time Sanibel resident, has killed "lots of Diamondbacks." The most recent one, about five years ago, was a five-and-a-half-foot specimen found at his home on Periwinkle Way. Opal Combs told me of one she saw at her home, Woodmere Preserve, in 1956. Dick Workman saw one caught by Allen Nave and later released in July of 1974. In that same month, ornithologist Richard Beebe found one at the end of the Causeway. In the spring of 1974, a young man of my acquaintance also caught one at the end of the Causeway and kept it alive for many months. Jean LeBuff told me of one seen crawling across the Causeway, and her husband, Charles, has seen many others. Jim Anholt watched a very large one cross Beach Road on Thanksgiving Day, 1975.

Perhaps the most interesting Rattlesnake anecdote that emerged from my inquiries around the Island was the one told me by Arthur Hunter, famed Hibiscus hybridizer, concerning his three-and-a-half year old female dachshund, "Toddy". One day back in the summer of 1971, she asked to be put out the back door of her home at Rabbit Road and Gulf Drive. Her master obliged and some time later went to the door to let Toddy back in. To his amazement he found that she had killed and eaten a Diamondback. All that remained was the head, tail and some skin. How this gentle little dachshund managed this can only be surmised. She is just about the cuddliest of dachshunds and for her to take on and actually kill and eat such a lethal opponent as a Diamondback is amazing indeed.

This Rattlesnake belongs to a family known as Viperidae, subfamily Crotalinae, the Pit Vipers, the nature of which will be explained later. Its technical name is *Crotalus adamanteus.* In trying to determine how large *Crotalus adamanteus* actually grows, one finds many references to a specimen that was eight feet nine inches long. Charles Bogert, formerly of the American Museum of Natural History, investigated this case and found that it was based on measuring a skin. One can't really measure a snake skin and get an accurate estimate of what the animal's size was in life because skins do stretch. The late L. M. Klauber of the San Diego Zoological Society, well known for his extensive work with Rattlesnakes, investigated the size of our Eastern Diamondback. He stated the maximum to be slightly over eight feet, and this is the statement that I accept.

The Crotalids have developed the most efficient venom delivering apparatus of any reptile living or dead. The fangs are injection needles — hollow, long, and capable of being folded back against the roof of the mouth when not in use.

As mentioned previously, Crotalids are Pit Vipers. They have facial pits which are essentially sensory organs. Each pit, lying between the eye and nostril, consists of two cavities separated by a thin membrane. The outer cavity is readily seen to be quite open; the inner one is concealed, except for a tiny hole just in front of the eye. Similar depressions occur on the lips of some boas and pythons. It has been determined that in both cases, these organs discriminate between heat and cold. Thus, the Rattlesnake is able to lie in ambush beside a rabbit run, strike and inject its venom in a passing rabbit, even in the dark, with unerring accuracy, and then follow the heat and odor track of the rabbit, overtake the dead or dying animal some distance away and devour it whole.

The forked tongue of the serpent is another little understood but very important sensory organ of this much-maligned but valuable animal. Many people feel that the tongue is a "stinger", but this is not true. Actually it's an instrument by which objects are "smelled." The tongue is flicked out. Its twin tips carry material particles and gaseous molecules from the air or minute bits from anything or any surface it touches. The two tips deposit the substances into two tiny cavities in the forward part of the roof of the mouth. These cavities, called Jacobson's organs, are in fact akin to olfactory rather than taste organs, even though the tongue plays an important part in the whole activity. This "smelling" process aids in tracking prey.

The Diamondback Rattlesnake has keen vision, but virtually no hearing, since snakes have no external ears. It is easy to set up an experiment with very nervous Rattlesnakes that are prone to rattle and show great disturbance at the slightest visual stimulus. These same animals will remain absolutely quiet when a loud sound is made near them.

The rattle consists of the shed skin of the terminal tail scale and serves to warn potential enemies. Each time the skin is cast, a new rattle segment is added. Of course it is untrue that the number of rattles indicates the number of years the beast has lived — especially when we consider the fact that an old Rattler may have lost some of the terminal, or first, rattles. The rattle count merely indicates the number of times the snake has shed his skin in his lifetime. Under excellent feeding conditions, a snake may shed every month or six weeks. Under very poor conditions, the snake may not cast his skin for many months.

The hemotoxic venom of the Eastern Diamondback Rattlesnake is not as lethal, drop for drop, as many other snake venoms. However, the reptile is so large that it can deliver a sufficient quantity with a single bite to make it a very dangerous animal indeed — one that demands respect and should never be taken lightly. If you should ever find one here on Sanibel, just leave it alone. Let it go its way because it, too, has its important place in our complex ecosystem. If you find one that is too close to home for comfort and would like to have it transported, any member of the Sanibel conservation community will gladly respond to your call.

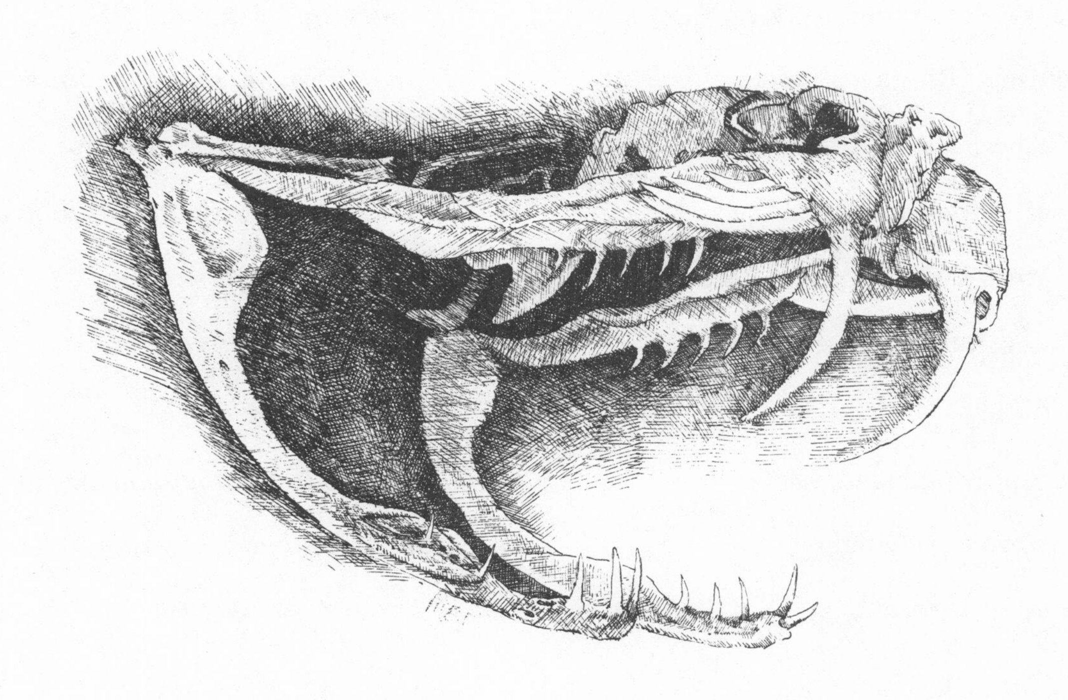

Skull of Eastern Diamondback Rattlesnake. Note replacement fangs.

CHECK LIST OF THE SNAKES OF SANIBEL

COMMON OR VULGAR NAME	*LATIN NAME*
Florida Water Snake	*Natrix fasciata pictiventris*
Mangrove or Flat-Tailed Water Snake	*Natrix fasciata compressicauda*
Florida Green Water Snake	*Natrix cyclopion floridana*
Florida Brown Snake	*Storeria dekayi victa*
Southern Ribbon Snake	*Thamnophis sauritis sakeni*
Eastern or Common Garter Snake	*Thamnophis sirtalis sirtalis*
Southern Ringneck Snake	*Diadophis punctatus punctatus*
Southern Black Racer	*Coluber constrictor* priapus*
Eastern Coachwhip Snake	*Masticophis flagellum flagellum*
Eastern Indigo Snake	*Drymarchon corais couperi*
Yellow Rat Snake	*Elaphe obsoleta quadrivittata*
Red Rat or Corn Snake	*Elaphe guttata*
Scarlet Kingsnake	*Lampropeltis triangulum elapsoides*
Eastern Coral Snake	*Micrurus fulvius fulvius*
Eastern Diamondback Rattlesnake	*Crotalus adamanteus*

***Does not constrict, in spite of its name.**

Eastern Diamondback Rattlesnake

Six-lined Racerunner

15 *THE ORIGINAL SANIBEL STREAKER* And Other Lizards

All of us have glimpsed the Six-lined Racerunner, for it is the original "Sanibel Streaker."

Few people ever see this little lizard close up unless one of them falls into a swimming pool or tries to hide under a moving lawn mower. But this very high-speed streaker can be seen almost daily on Sanibel as it escapes from one's path almost anywhere that is dry and sunny.

This interesting creature runs for short periods of time at speeds that have been estimated to exceed 20 mph. You have to be very nimble indeed to capture one of these lizards alive and unhurt. And not the least of the difficulties is his habit of stopping abruptly while your eye continues to move on, so you lose track of where he really is. To see one close up, the best thing to do is to wait for one of these frigid mornings that we have so few of, and turn over a few logs or stones until you find one chilled and resting motionless, waiting for the sun's warmth to render it active again.

A chilled streaker can't run very fast and you will have almost a second to grab him. But be careful, for his tail will very readily break off in your hand. In common with many other lizards, he utilizes this break-away tail as a major defensive mechanism. When it breaks off in a struggle, the tail wriggles frantically on the ground, attracting the attention of the predator while the recent owner makes his successful escape.

Immediately after this, the tailless creature starts to grow a new one and in a matter of months he is just as good as new, except that the new tail may be just a tiny bit shorter with a scar where the original tail was broken off. An original tail, only partly broken off, may stimulate the growth of a new one, resulting in a fork-tailed animal.

This little streaker is a very valuable animal in our island habitat because it feeds mainly on insects and consumes inordinate numbers of them. It likes cockroaches (which some residents prefer to call, euphemistically, "palmetto bugs"). The Racerunner also eats beetles, mosquitoes and other insects such as flies, grasshoppers and crickets.

Our little *Cnemidophorus,* which is his generic name, is also a valuable ecological measuring tool. It measures the extent of the habitat destruction of our island. How this is interpreted may be of interest. The Racerunner prefers what the zoologists call a "xeric" situation — that is, a semi-desert situation. If you modify a *Spartina* marsh with a few feet of land fill so that it is no longer wetland, then you create a xeric or desert condition which is more favorable to the Racerunner than the original dampness. Consequently, while some other animals are declining in their abundance on Sanibel due to habitat destruction through loss of wetlands, the Six-Lined Racerunner is becoming more abundant. It occurs to me that developers and lizards make strange bedfellows.

Racerunners belong to a group known as the Teiids. There are seven species in the United States, but only one east of the Mississippi — our Six-lined Racerunner, *Cnemidophorus sexlineatus* whose specific name, of course, refers to the six lines on its back.

There are two hundred members of this family, Teiidae, most of which are concentrated in the American Tropics. None lives outside of the Americas, making this the most characteristic lizard family of the New World.

In the West Indies and in the Bahamas, one sees Racerunner-like lizards, but these, the *Ameivas,* are somewhat different from ours, and some grow much larger. The largest Teiid of all, the Tegu, *Tupinambas,* grows to be over three feet long and is very destructive to chicks and eggs in areas of South America where the poultry industry is important. This egg thief is very speedy and at the slightest disturbance will go crashing off through the underbrush, making a sound more like that of a wild horse than a ten pound lizard. I recall chasing these on horseback in Mato Grosso, in the interior of Brazil, some years ago and unsuccessfully trying to "bulldog" them, a skill my gaucho companion had mastered.

Another Teiid is the Caiman Lizard, *Dracaena,* of northern South America, which lives in the water like the Alligator or the Caiman and has scales and scutes that are very like those of the Alligator. Its food is largely aquatic, principally consisting of snails. It crunches the shells with its rounded pearl-like teeth, swallowing the soft bodies and spitting out the sharp shell fragments.

Our Racerunner may grow to be twelve inches long and has a long and rather pointed head covered with large shields. It has strong hind legs and a long tail. There are very fine granular scales on most of its back, and the ground color is brownish with the stripes and spots yellow. The males are brilliantly colored on the chest, throat and belly with bright blues and some orange and pink.

In addition to the Racerunner, there are eight other species of lizards on this island, three of which are exotic introductions.

One of the most fascinating that has been observed over the last couple of years on Sanibel Island is the largest lizard of the New World, a species common to much of South America and Central America as far north as Guatemala. It was often used in "monster" movies where it drew shrill screams at the Saturday matinees. I refer to the Green Iguana that attains a length of more than six and one half feet. The accompanying drawing was made from life, of a four-foot specimen that was captured near the post office by Jessie Dugger in March of 1975.

Others exist here too: Grace Whitehead has her own pet, "Georges," which she has been known to wear to parties much as others might display an emerald jewel. Georges now lives free in her backyard and has grown too large for handling.

Other Green Iguanas have been seen on Gulf Drive and at "The Rocks." Charles LeBuff captured one on June 6, 1974, at the eastern end of the Island. Dick Workman saw one in the talons of a hawk. "Green—too big for an anole." Three live at the home of cinema magnate Miriam Stone where her niece, Pam, tamed them in the summer of 1976 to the extent that they would come and feed from her hand.

Hundreds of thousands of Iguanas were imported into Miami and Tampa for the pet trade and have been distributed everywhere, usually as tiny babies. Thousands have escaped into the environment, most of them fated to die because of the rigors of our northern winters. But many of those that escaped in Florida have survived and this species is rather well-established in this state. They can be seen frequently in Dade and Broward Counties and now in Lee County. A four-foot specimen was killed in September 1974 on a road in Fort Myers. I cannot be sure that they are actually breeding on Sanibel, but it seems to be so, as juveniles are seen.

The Iguana, whose technical name is the same, squared, *(Iguana iguana)* is a member of a very large family of mostly New World lizards. The only lizards in the whole world that are larger than the Iguanas are the Monitor Lizards of Africa, Asia and Australasia — primitive animals, probably rather like those early lizards of past ages that were the ancestors of modern serpents.

The iguanid lizards are equipped with a pineal body or a third eye on the top of its head. This ancient ancestral organ is probably light-sensitive, although it is hard to determine for sure, for communication with Iguanas is not easy.

The Iguana uses its tail as a weapon. A cornered Iguana will lash out brutally with its tail as it gapes its mouth which is equipped with formidable sharp teeth capable of biting your hand clean to the bone. This is largely a bluff, however, for even a large Iguana will escape with great rapidity if given half a chance.

The Green Iguana lays eggs which are consumed by many Latin Americans. Adult Iguanas are preyed upon by man for the pet trade and for their eggs and meat. Consequently this species sometimes called the "Common Iguana" is not as common as it used to be.

Thirty years ago, I remember seeing in the market at Managua, Nicaragua, a large pile of live Iguanas. They had the long toes of the hind foot cruelly cut and the tendons pulled out and tied in knots across the back, thus immobilizing them. They were kept alive in this condition until sold. Some of them weighed seventeen or eighteen pounds. Recently I returned to the Managua market and saw only a few tiny, skinny little dressed out Iguanas. I think this is a rather good indicator of what has happened to this species in overpopulated, protein-starved Nicaragua.

I don't want to leave this subject without mentioning a few other iguanids. Remember the small chained green or brown lizards that used to be sold at the circus? They were the same species we see around our homes. This animal, the Green Anole, is native to Sanibel as well as to most of the southeastern part of the United States and many of the nearby Bahama Islands. It is an intriguing little creature in that it changes from dark brown to light green with many shades in between. It even assumes strange colors such as purple just before death, or after a serious injury.

Male Anoles are quite defensive of their territories and frequently may be seen quarreling with each other by displaying the dewlap, that pinkish gular appendage which extends downward from the throat and which is conspicuously extended and bobbed by action of the head. Two males may display until one disappears or is driven from the territory.

Eggs smaller than a pea are laid in loose humus or soil in damp shade. By late summer or early fall, tiny juveniles may be seen around porches, gardens, vacant lots and in all four tracts of the "Ding" Darling Refuge, feeding on the small soft-bodied insects that are so abundant at that time of year.

The technical name of this little creature is *Anolis carolinensis.* It is commonly called the Green Anole or false Chameleon — false because there are true Chameleons in Africa and India which are not at all like these small iguanid lizards.

Green Iguana

There is another Anole here on Sanibel, an invading interloper, commonly known as the Key West Anole, *Anolis sagrei.* Key West was just an intermediate stop for this lizard on its journey to Florida from the Bahama Islands, where it is indigenous.

As an old resident of Eleuthera, Bahamas, I have long acquaintance with the "Key West" Anole, it being the most populous of four Anole forms native to that island. It is definitely larger and its color-changing abilities are somewhat less developed than those of the first-mentioned species. It is only able to change from light gray to deep dark brown, seldom showing any other colors except occasional yellow, but never green. There are sometimes chevron markings on the backs of males and these become evident when they are posturing or fighting off other males. The female shows similar dorsal markings. When defending their territory, their behavior includes an impressive extension of the colorful dewlap, or throat pouch, and an exhibition of vigorous "pushups."

These Anoles probably came to Sanibel in potted plants from Dade County, where they are very common animals, having arrived there from the Keys about four decades ago. It is not known for sure whether or not this exotic animal is having a deleterious effect on our ecosystem, but if not, it will prove to be somewhat an exception; for a phenomenon we see ever more frequently on Sanibel among both floral and faunal communities is the disruption of natural systems by introduced exotics. The very silhouette of Sanibel is exotic, the Australian *Casuarina* being the most evident of all plants. The nature trails through the Darling Refuge and Conservation Foundation Tract are overpopulated with unbelievable numbers of Brazilian Pepper Trees. A predatory frog species which is growing in abundance is an exotic from Cuba. Australian Budgerigars ("Budgies") are breeding on the Island. A pair of Indian Hill Mynah birds live at "The Rocks." Two monkeys have been seen on the main Sanctuary tract; another was run over on Tarpon Bay Road.

Most exotic introductions prove to be damaging to the environment. Certainly the House Sparrow and the European Starling and numerous other birds have caused great havoc. Here on Sanibel we talk about preserving the original state of our natural systems, but actually a great many exotics already live here among us so that what we see today is quite unlike the island Ponce de Leon visited. This unfortunate situation is largely uncorrectable.

The horned "toads" of southwest United States are iguanids. These lizards feed largely on ants and with the ant explosion on Sanibel one might have wished that these horned toads had found their way to our Island.

Another iguanid not found on Sanibel is the Basilisk. Basilisks, like some dinosaurs of old, can walk and run on their hind legs, dragging their tails behind them. But perhaps the most startling thing about the Basilisk is that it can dash across the surface of water without breaking the surface tension. It is known throughout Latin America as the Jesus Christ Lizard because of this "miraculous" ability.

One of the most remarkable reptiles is the Galapagos Island Marine Iguana, the only lizard in the world that has adapted to the marine environment. It rests on the rocky beaches of the Galapagos most of the time when not feeding on seaweed under the surface of the waters of the surrounding sea.

In the Caribbean there are many species of iguanas, each a little different from the other, having been isolated for eons on separate islands. Recently I made an expedition to the Caicos Islands north of Haiti to observe one of these seriously endangered animals. There I found the remaining populations seriously threatened, for it is being eaten by hungry natives whose diets are seriously lacking in protein. There are five or six forms in the Bahamas and others in Cuba, Puerto Rico, Haiti, Grand Cayman and elsewhere in the Caribbean.

Darwin need not have taken that long dangerous voyage around the Horn on HMS BEAGLE to the Galapagos to gather the data that finally resulted in his Theory of Evolution. He could have spent those months cruising among the Caribbean Islands studying the diversity that had evolved in the Iguana genus known as *Cyclura.*

I hope that introduction of the South American Iguana does not prove to be a menace in the areas where it seems to be establishing itself here in Florida. I really can't conceive of any problems that it can cause beyond the fact that a population of them could devastate a lettuce field. What they might do to a strawberry patch is

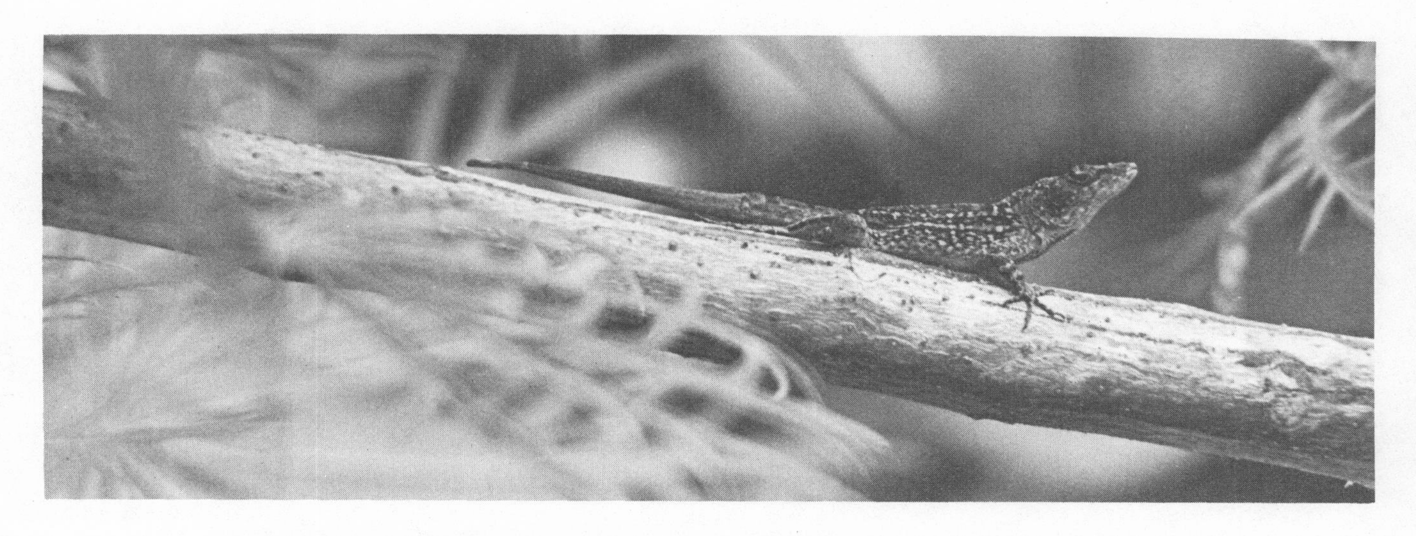

The "Key West" Anole. This Bahamian exotic is very common on Sanibel.

Photos, Dallas Kinney, *Island Reporter*

The Green Anole is native to Sanibel. Here two are shown mating.

Photo, Laura Riley

Southeastern Five-lined Skink. The tail is easily broken off in a struggle with a predator. Its wriggling attracts attention while its original owner runs off to prosper and grow a new tail.

unimaginable. But on Sanibel, where agriculture is not of much significance, I don't believe this great lizard is going to do any harm beyond eating *Hibiscus* flowers which, after all, are renewed daily.

Many people on Sanibel have expressed interest in viewing one of these striking animals. They are beautiful, often a bright pea green, sometimes displaying the dark markings as shown in the drawing. They do change colors, and sometimes assume a much darker hue.

I, for one, shall be interested in watching the progress of this interesting introduction, and when I see one on the road I shall slow my car so as not to flatten it.

Before winding up this chapter, I want to make sure to mention that fascinating exotic from India, the House Gecko, *Hemidactylus garnoti,* also called the Indo-Pacific Gecko. This species undoubtedly was brought to the American shores on cargo from India and established itself in Miami. It being somewhat of a domestic form (that is, it lives around human habitation), it is not strange that its center of distribution on Sanibel Island would be the Periwinkle Way Trailer Park. Trailers have been brought across the state from Miami for a long time. I therefore postulate that the species has become established on Sanibel by being thus transported. It's rather a nondescript little four to five-inch long delicate-skinned brownish grey gecko with small rounded dorsal scales and large toe pads. Its back is almost a uniform marble color with dark brown, sometimes with small whitish patches and it has no eyelids. It can be found on the sides of buildings, is nocturnal, and feeds on insects, so it's hard to believe that this species can in any way be a detriment to our island environment.

But the incredibly fascinating thing about this lizard is that it is unisexual; that is, it occurs here only as a female population. There are no males at all. Yet females lay eggs, the eggs hatch, and new little female lizards are produced. This process is called parthenogenesis. Some of the lizards related to the Racerunner demonstrate this same phenomenon. I find it interesting, indeed, but personally am glad that it is a phenomenon not encountered in the human species.

Two other common lizard species of Sanibel are the Southeastern Five-lined Skink, the adult males of which have reddish heads. These lizards are five to seven inches long. The juveniles have bright blue tails and are somewhat smaller.

Our other skink is the Ground Skink, which is very thin with tiny almost vestigial legs so that it progresses as much by wriggling as it does by running. It is small, seldom reaching five inches in length. It has a smooth brown back, sometimes golden brown, with dark stripes on the side. It is unique in having in the lower eyelid a transparent disc through which the lizard may see even when its eye is closed.

This animal, as well as the other Skink, can be seen in most gardens, yards and vacant lots on Sanibel.

One other interesting, and extremely rare, lizard here on Sanibel Island is the Eastern Glass Lizard, which has no legs at all, and is most fragile. One of the two specimens that have been seen was found under *Casuarina* duff near Bowman's Beach. This form is not at all uncommon on the mainland, but on Sanibel it is very rare.

Following is a checklist of lizards of Sanibel Island.

CHECKLIST OF THE LIZARDS OF SANIBEL

VULGAR NAME	*LATIN NAME*
Indian House Gecko	*Hemidactylus garnoti*
Common or Green Iguana	*Iguana iguana*
Green Anole	*Anolis carolinensis carolinensis*
Key West Anole	*Anolis sagrei*
Florida Scrub Lizard	*Sceloporus woodi*
Southeastern Five-lined Skink	*Eumeces inexpectatus*
Ground Skink	*Leiolopisma laterale*
Six-lined Racerunner	*Cnemidophorus sexlineatus*
Eastern Glass Lizard	*Ophisaurus ventralis*

16 THE TURTLES AND TORTOISES OF SANIBEL

Sanibel is rich in turtle and tortoise species. In addition to five marine forms, the land and fresh waters of Sanibel nurture another eleven kinds. The most striking are, of course, the huge marine turtles, the Loggerhead being the one most often seen. These great animals grow to be thirty to forty-five inches long, the record being a forty-eight-inch turtle. Their weights range from one hundred seventy to three hundred fifty pounds. One old monster weighed in at five hundred pounds!

The conservation of this species is the subject of a fine Sanibel-based conservation organization known as Caretta Research, Inc., *Caretta* being the technical name of the Loggerhead Turtle. Many papers and studies originating from that organization are available and its president, Charles LeBuff, is always happy to discuss the problems of the Loggerhead.

Caretta breeds on Sanibel's beaches. Every year there are quite a number of nests, although this number is declining. The organization carries out beach patrols each night during the breeding season, from May to August, and many valuable data are collected. Sometimes eggs that were deposited in unfavorable nesting locations are gathered and artificially hatched for eventual release.

Loggerheads do not usually travel the open ocean directly to their nesting destinations, so the females laying on Sanibel's beaches probably come from the Gulf and the not too distant bays and sounds. By May the turtles have arrived in the vicinity of the nesting beach, having paralleled the coastline perhaps for miles. Both sexes are present, close to the shore, but only the females approach the land and actually crawl out of the water. The males spend their entire lives at sea.

Mating takes place in the water, often some distance from the shore. There may be sperm retention in the Loggerhead; that is, copulation in the summer of 1977 may fertilize eggs to be deposited in the summer of 1979. The female seeks a proper location to deposit her eggs and begins her slow ascent of the beach. Sometimes she turns around and reenters the sea. This activity is popularly known as a "false crawl." Later the same night, or perhaps after several nights, she will again approach the beach. The nest will be excavated and the eggs deposited. Construction of the nest is accomplished by the use of the hind flippers after which the eggs are dropped into the pear-shaped cavity. There can be anywhere from seventy-five to two hundred eggs in a single nest — usually more than one hundred. The egg is white, the shell is a bit flexible, and about the size of a pingpong ball. After some sixty days, the young turtles will emerge at night from the nest and, conditions being right, will head for the water. Approximately one quarter of one percent will achieve sexual maturity. The rest will be victims of various predators.

The predators of eggs and hatchlings are many and can be divided into two groups: natural predators such as raccoons, ghost crabs, fish and birds, and non-gourmet type human beings who find the flesh and the eggs edible. A third major problem is weather damage. High storm tides and chronic beach erosion can wreak havoc with turtle nesting sites. The raccoon used to be a very significant Loggerhead Turtle predator on Sanibel, but with the increased availability of refuse and handouts from human residents, the raccoon has found an easier food supply and as a consequence, fewer Loggerhead Turtle nests are disturbed by coons these days. Also, Caretta Research workers use a quaint but effective method to discourage coons. After a mother turtle has laid her eggs and returned to the sea, the workers camouflage the tracks and nest and then urinate on the site. Human urine effectively disguises the nest site and reduces coon predation.

The actual hatching of the eggs takes place several days before the young Loggerheads appear at the surface. In common with most other turtles, the young Loggerheads are equipped with an egg tooth for cutting the shell. After they emerge from the shells, reaching the surface is a cooperative effort. Many turtles emerge from the shells and move *en masse;* sand is loosened above and trickles down so that in fact the animals are raised on a kind of elevator. With the mass of wriggling life a foot or two below the surface causing sand from the top of the cavern to fall to the bottom, the floor is actually raised and in due course the youngsters emerge onto the surface of the beach.

If their activity takes them to a layer of sun-heated sand, a response is triggered telling them that it is day-time and the whole mass quiets down until evening when the activity recommences and the babies emerge at the surface. They seldom come up in the daylight hours. Sometimes this emergence can take as long as a week until the last animal has reached the surface. Normally the two-inch youngsters, upon emergence, head right for the water which they enter with great activity and swim straight out to sea. The Loggerhead Turtle, as it emerges, responds to a positive phototropic mechanism that causes it to head for the lightest part of the horizon. The primeval beach, before the advent of condominiums, beach lights, automobiles, parking lots, etc., provided the brightest light area directly toward the sea. Consequently, the ancient urge of the Loggerhead was successfully satisfied when it headed for the sea under those pre-human conditions. Today, the beach at Sanibel provides a lethal trap, with artificially lighted areas spread from the Lighthouse end of the beach west to Bowman's Beach. From that point on, the problem is a lesser one due to lack of residential development. In 1975, one hundred and four baby turtles were found wandering around the parking lot of a building near mid-Island. Fortunately they were captured and released in the sea.

One Caretta Research worker, Paul Zajicek, developed a unique system to help disoriented baby sea turtles to head for the water. He simply stands out in the surf with a bright torchlight and attracts the youngsters with this so that they immediately crawl toward the sea.

Another new problem facing the Sea Turtles of Sanibel are sea walls. So much of the area around the condominium concentration is seawalled that it is almost impossible for normal egg-laying to take place. Sometimes a female Loggerhead, confronted with a seawall and with the urgent necessity of depositing a clutch of eggs, will actually oviposit in very inadequate nest sites with little or no prospect of successful hatching.

Many valuable data are emerging from the work of Caretta Research. One of the most interesting contributions to the study of this species is the discovery of multiple nesting. Up to six nests from one female in a single laying year have been documented. Usually these turtles will not lay in successive years, however, and the group laying this year will not be seen on the beach again until year after next.

The laying of Loggerhead Turtles on Sanibel Island has been on the decline for many years. Sanibel stands to lose this valuable species unless proper, adequate, remedial measures can be taken.

For those interested in more information on this fascinating and valuable species, I suggest that the publications and studies of Caretta Research Inc. be obtained.

The other Sea Turtles here are so rare as to be almost unheard of, except for the Green Sea Turtle which exists as a population of subadults in the Bay areas of Sanibel waters during the summertime.

The Great Leatherback, the Atlantic Ridley and the Hawksbill are known to exist in the area, but are almost never seen.

The Gopher Tortoise, *Gopherus polyphemus,* is a familiar animal here on Sanibel. This is the last true land tortoise (a term generally applied to strictly terrestrial chelonians) to inhabit this part of the world.

But not so long ago, geologically speaking, there were very large tortoises in Florida. In the mid-tertiary there was one that attained more than six feet in length with a shell that was almost two inches thick.

Tortoises in general have encountered hard times in recent millenia so that today their numbers are greatly reduced throughout the world. Thus we are indeed fortunate to have this fine species here on Sanibel.

Photo, Caretta Research

Charles LeBuff with a specimen of his favorite species, the Loggerhead Turtle.

Our Gopher attains a maximum length of fourteen and a half inches, is a heavy bodied beast and, unfortunately for him, has been consumed by mankind over recent centuries, first by the Indians of this region and more recently by local Floridians.

Eating Gopher is a little bit like playing Russian Roulette, however, because they do sometimes consume toxic mushrooms. This does not seem to hurt the tortoise, but it can be lethal to the person who elects to eat Gopher stew for dinner. Even if I weren't an avid fan of the Gopher and one of his most enthusiastic conservators, I would never take the chance of eating his flesh.

The Gopher Tortoise digs a long burrow. The flattened forearms are adapted to digging burrows which may attain a length of over forty feet and a depth of over ten feet under the surface of the ground on the mainland but in a barrier island habitat such as Sanibel's, a depth to just above the water table is about the limit.

These burrows are shared by many other animals, making the ecology of the Gopher hole rather interesting and important. Opossums, rabbits and some rodents all share the burrow with the tortoise, and it is thought that they live peaceably together. One frog, appropriately called the Gopher Frog, has evolved with the Gopher, whose burrow is the frog's exclusive refuge when not breeding in a nearby freshwater pond. This frog lives on the mainland and is not counted among our species.

The most startling inhabitant of the Gopher hole is that highly specialized engine of destruction, the great Eastern Diamondback Rattlesnake.

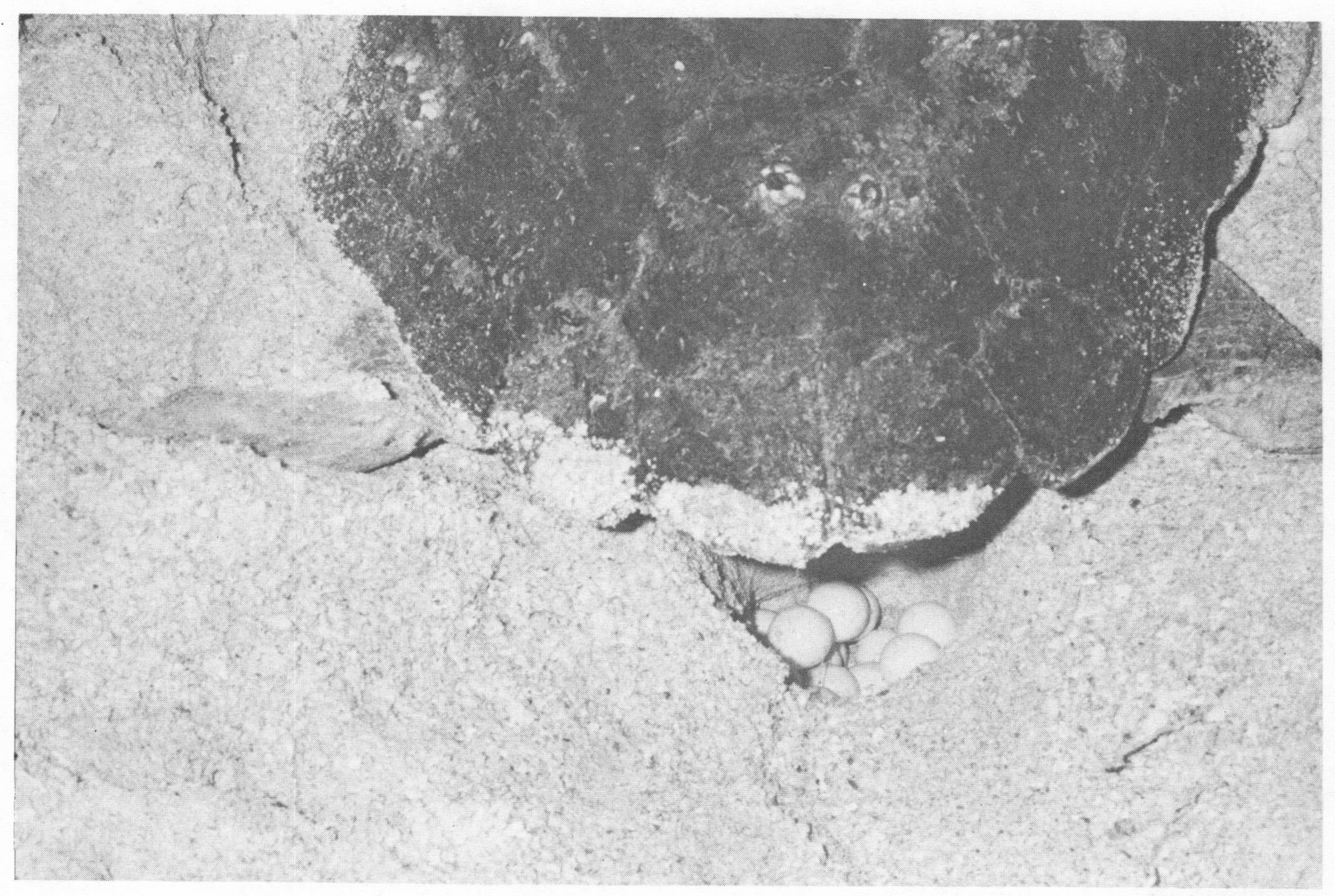

Photo, Charles LeBuff

Loggerhead Turtle laying her eggs

Elise Fuller of Palm Lake Drive had a unique opportunity to contribute to the knowledge of Gopher Tortoises. In July 1974, a small eight-inch Gopher moved into the flower bed in her front garden. It dug a proper hole and lived there until July 1976 when it took off for parts unknown.

Mrs. Fuller named this animal Beelzebub, which might seem to be a rather harsh name for such a charming creature. She observed and recorded his daily actions for the whole of 1975 as well as for part of 1974 and 1976. She took temperature records, timed the animal's first appearance in the morning and recorded the weather conditions of the day. Her records reveal that the animal usually came out at ten or eleven in the morning and only when the sun was out and the entrance to his burrow had warmed up. On cold days, he did not come out at all, nor did he appear on rainy days. In the summer and fall he often appeared in the afternoon to feed, but in the wintertime this activity stopped.

Beelzebub was fed high protein dog food and vitamins, together with other more normal foods such as flowers — especially Hibiscus — and some grasses. On such a fine diet, Beelzebub grew to be a fine specimen.

Beelzebub must have built an attractive burrow, for a Florida Black Racer, *Coluber constrictor priapus,* moved in with him and these two animals shared the burrow peaceably for a period. Aside from the pleasure Mrs. Fuller derived from Beelzebub's living at the front door, she was in a position to make some unique contributions to the knowledge of this species. Her data have been referred to Dr. Peter Pritchard, a world authority on turtles and tortoises.

For a reptile, the Gopher is a very intelligent

animal. It can learn to recognize individual people and voices and will learn rapidly to come to friendly persons, eat from the hand or eat from a feeding tray much as birds do.

Gopher holes are being bulldozed in many areas on Sanibel where development is taking place. The Gopher is able to dig a long burrow and move many hundreds of pounds of gravel or soil, but only when he is working from top to bottom. In front of each Gopher hole you will see a spoil pile consisting of perhaps two tons of sand and shell. But when a bulldozer crushes the hole above him, he is unable to move this compacted soil and consequently dies a slow death.

In order to save as many of these animals as possible Richard Workman, Director of the Sanibel-Captiva Conservation Foundation, and this writer have endeavored to rescue some from threatened land and place them in safe habitat.

A principal partner on our rescue team is my dog, Boris, who can sniff a Gopher hole and immediately tell you by gesture whether it is inhabited. If it is, he can dig one out in a few minutes, expending much frenzied energy. But Boris's activity must be carefully supervised, else he will bite through the not very thick shell and injure the beast. This he recently did, due to my carelessness, and the specimen, a small six inch example, was bleeding from several deep puncture wounds in his carapace. Dick Beebe was persuaded to take on this case, and with careful nursing and good food, the animal recovered and was released in the Gopher reserve which is maintained on part of the Conservation Foundation's land.

A careful estimate of the numbers of Gopher Tortoises destroyed in 1974 by the bulldozing along Wulfert Road suggests that we lost more than two hundred animals in about two weeks.

The Gopher Tortoise commonly buries its eggs near the spoil mound near the entrance to its burrow. The eggs are spherical in shape, with brittle shells, and about an inch and a half in diameter. They are deposited in May, June or July and hatch in about 60 days. Usually there are only about a half-dozen and the youngsters disappear and are seldom seen until they are five or six inches long. They grow rather rapidly. One six-month growth record chalked up a full inch.

There are three other species of *Gopherus* in the world — one in the far west, one in Texas, and *G. flavomarginatus,* which amazingly was only discovered in 1959. It is confined to three of the northern states of Mexico. It is really outstanding because it attains a great size for a Gopher, over three feet in length.

There are still many species of tortoises in the world, but most of them are under great pressures. Two species that are perhaps most interesting are two insular forms, one inhabiting Aldabra, a British island in the Indian Ocean just north of Madagascar, and the other inhabiting those Ecuadorian islands made famous by Charles Darwin, the Galapagos, lying in the Pacific several hundred miles off the coast of South America.

The Aldabra tortoise has fared rather well, all things considered. Aldabra has been set aside by the British as a nature reserve and these animals are breeding abundantly under the careful protection of British conservationists.

The Galapagos form in its several races, however, fared badly at the hands of man, especially during the last century when the whalers, on rounding the Horn on their way to whaling grounds in the Pacific, would stop at the Galapagos Islands to load their holds with live tortoises. They would turn them on their backs and gradually before the animals starved to death, would consume them and thus have good fresh meat for long voyages that sometimes lasted three years. So the Nantucketers not only contributed to the demise of the great whales, but also brought untold destruction to the great tortoises of the Galapagos.

The International Union for the Conservation of Nature has declared the giant tortoise of the Galapagos to be an endangered species, as have our government and many governments of the world. There is the Darwin Biological Station in the Galapagos now and efforts are being made to bring these animals back from the brink of extinction. The San Diego Zoo is also successful in breeding them, so this animal probably will not disappear and future generations will be able to marvel at this great living relic of the past.

Another well-known but diminishing turtle species of Sanibel is the Florida Box Turtle. According to LeBuff, this animal was very common on Sanibel in years gone by. But automobile traffic has taken its toll, and I believe it to be rare today.

Gopher Tortoise

Between May and October, 1974, with diligent searching, I located nine specimens. Close observation during the entire year of 1976 turned up no specimens.

This species lives in the central fresh water slough and if it's to be found anywhere, the Sanibel Captiva Conservation Foundation wetland area would be the place to look. It does, however, invade other habitats, as is illustrated by a most interesting record provided by Mrs. Charles Fishburne, who found a large handsome male in the Gulf surf. It washed ashore and crawled back to more normal habitat.

An interesting occurrence took place on Sanibel not long ago in the home of ornithologist Dick Beebe. Beebe, who sometimes branches out into herpetology, has for a long time kept a number of box turtles on his patio among the potted plants. One of them, a Florida Box Turtle, demonstrated the unmistakable restlessness of a female turtle seeking an egg-laying site. Beebe provided this by placing an inclined plane from the floor up to the top of a four gallon planter which contained loose damp soil. Thomasina, who is a rather bright creature, found this board, walked up the inclined plane and laid her eggs in the soil. Approximately two months later, two babies hatched.

This is a very unusual occurrence because Box Turtle youngsters are seldom seen, being very secretive and of a more aquatic nature than the adults. It is not at all common to have a pair of young Box Turtles to observe and this incident was a rare treat indeed.

Box Turtles are unusual in many ways. Like the Gopher, they are heavy feeders on fungi including poisonous mushrooms and toadstools — the toxins of which don't seem to poison the turtles. But every so often there are a few unsuspecting people who think this animal might make a tasty stew or soup. Such misguided people themselves may suffer the agony of mushroom poisoning. Besides fungi, Box Turtles also eat slugs, snails, insects, earthworms, carrion, and many types of vegetable material, especially soft fruits. Here on Sanibel, wild guavas and papayas are popular.

As nearly everybody knows, Box Turtles get their name from their ability to close up the shell like a box. The plastron, or lower shell, is hinged front and back and can close tightly against the carapace, or upper shell, with the feet, head, neck and tail neatly tucked away inside, safe from predators. Normally a closed up Box Turtle offers only a hard surface to the outside world. But occasionally a gluttonous one will overeat and due to internal pressures be unable to close up both fore and aft. Thus he will leave either his head or tail exposed, so when danger comes he may be in trouble.

The plastron is not hinged during the early years of the Box Turtle's life and only develops to full functional use when the animal is about half grown. There aren't any great external differences between sexes in the Box Turtle, but the male does display a smooth concave depression in the posterior half of the plastron which aids in copulation.

The Box Turtle is among the longest-lived of all the reptiles. There are valid records which show the life of some to exceed fifty years. Florida herpetologist and leading turtle authority, Peter Pritchard, has accepted a longevity figure of one hundred twenty-three years for a Box Turtle. If true, this certainly has to be the longest recorded lifespan for any reptile.

My opinion of the Box Turtle coincides with that of herpetologist Clifford Pope who calls it "a quaint element of our fauna." It is certainly worth preserving as an altogether attractive animal. Especially, it should be cared for here on Sanibel where it is already rare.

Sanibel has its destructive hobbyists, however, who like to hear the sound of the animal exploding under the wheels of a car. In 1975 I pulled a turtle virtually from under the wheels of a truck in front of the Dairy Queen. The week before that, one was squashed on Periwinkle Way just in front of the Conservation Foundation.

Archie Carr, one of America's leading champions of the turtle, has written, "There exists a curious number of witless or psychopathic characters who love to run over box turtles just to hear them pop, and there is probably nothing much that can be done about these people except to hope they skid."[1]

We will not discuss each turtle species that lives on Sanibel, but at the end of this chapter there will be found a check list of them all. I do, however, want to mention one exceptional form that occurs here in considerable abundance. This is the Florida Softshell Turtle.

In the springtime we see an abundance of large, 18-inch round flat turtles on the roads and in our yards. They are out of their normal element which is fresh water, whether it be sloughs, lakes or ponds. At other times of the year the only time we see them is when one unfortunately bites the hook of a fisherman's line, creating a problem of trying to save the animal.

The reason the large females are abroad on dry land in the spring is that they are seeking nesting sites. Drivers are urged to be careful when spotting them on the road where too many have met premature deaths. I have counted over twenty dead ones in one season and have removed many live ones to safe habitat away from traffic. If you should try to save a threatened Softshell, it is wise to be very cautious because this creature has a long and agile neck and his pointy head is armed with powerful, knifelike cutting jaws. A big Softshell can deliver a nasty bite. It is safe, however, to pick one up by using both hands and grasping the rear quarter of the carapace on both sides. Then one cannot be bitten, although it is still very difficult to hold the beast because her hind legs will be kicking and her leathery shell is slippery.

If you should see a Softshell on the road, it is suggested that you try to save it — not by catching it, but by "pointing" it off the road towards a roadside ditch, away from traffic. This can be done with your foot — or if you don't want to take even that chance, by the use of any available stick.

I feel that the preservation of Sanibel's wildlife is of sufficient importance that I do not hesitate to stop traffic in both directions while undertaking such rescue efforts. Most motorists understand and accept a short delay. Only a few become angry and abusive. For them I suggest that another island would be more appropriate — Manhattan, for example.

---

[1]Quoted by permission

Those females that are successful in finding a nesting site will choose a place that will not be flooded. Then, using the ancient technique common to all turtles, whether they be Loggerheads, Box or Softshell Turtles, they will skillfully hollow out a cavity, using only the hind legs, and will deposit from ten to forty eggs in this hole. Afterwards, they will carefully smooth the surface and leave without giving it further attention. In about sixty days the eggs will hatch and the youngsters will head for the nearest fresh water.

The Softshell belongs to a group that is abundant in North America, Asia and Africa, there being some two dozen species.

Another Turtle that occurs here is the Ornate Diamondback Terrapin. It is found generally in salt or brackish water, usually among the mangroves, although occasionally juveniles are cast up on the Gulf Beach. This is the animal made famous by the gourmets of New York in the 1890's, when Diamond Jim Brady created a demand for its flesh which

Florida Box Turtles, the male showing amorous intentions

drove the price up to $100 a pound. More recently it, or its near relatives, have been captive-bred and a not too successful turtle industry has resulted. Our Diamondbacks can be seen in the "Ding" Darling Refuge on the Memorial Drive, sometimes sunning themselves on the bank, sometimes trying to dig nest holes in the hardpacked road itself.

The Florida Snapping Turtle or Osceola Snapper is a rare animal on Sanibel in the fresh water sloughs, but it does occur and can be seen — sometimes on Island Inn Road, crossing from one fresh water bog to another.

There are several hard shelled turtles on the Island: the Florida Red-bellied Turtle, the Yellow-bellied Turtle, the Peninsular Cooter, (which is a corruption of a name of African origin — *kuta,* meaning turtle — brought to this country by early slaves), and the Florida Chicken Turtle all are relatively abundant in the Sanibel River, canals and real estate lakes. Successful breeding takes place every year and they are not at all hard to see.

The four hard shelled turtles all are native to this region except for the Yellow Bellied Terrapin which did not exist on Sanibel before 1964. In that year, a graduate student, E. Leach, released twenty-three specimens in the fresh water area about three miles west of the Lighthouse. Quite a number of specimens have been seen since then and this alien from North Florida seems to be surviving and breeding in this area which is not part of its natural range.

Following is a list of all the Testudines — the turtles and tortoises of Sanibel and the surrounding waters.

CHECK LIST OF THE TURTLES AND TORTOISES OF SANIBEL AND SURROUNDING WATERS

COMMON OR VULGAR NAME	*LATIN NAME*
SEA TURTLES	
Atlantic Loggerhead	*Caretta caretta caretta*
Atlantic Hawksbill	*Eretmochelys imbricata imbricata*
Atlantic Ridley	*Lepidochelys kempi*
Atlantic Leatherback	*Dermochelys coriacea coriacea*
Atlantic Green Turtle	*Chelonia mydas mydas*
BRACKISH WATER TURTLES	
Ornate Diamondback Terrapin	*Malaclemys terrapin macrospilota*
LAND TORTOISE	
Gopher Tortoise	*Gopherus polyphemus*
FRESH WATER TURTLES	
Florida Snapping Turtle	*Chelydra serpentina osceola*
Striped Mud Turtle	*Kinosternon bauri palmarum*
Florida Mud Turtle	*Kinosternon subrubrum steindachneri*
Florida Box Turtle	*Terrapene carolina bauri*
Florida Red-bellied Turtle	*Chrysemys nelsoni*
Yellow-bellied Turtle	*Chrysemys scripta scripta*
Peninsular Cooter	*Chrysemys floridana peninsularis*
Florida Chicken Turtle	*Deirochelys reticularia chrysea*
Florida Softshell Turtle	*Trionyx ferox*

17 *SANIBEL'S BOUNTY* Edible Plants

If some disaster should occur and you found yourself stranded on Sanibel Island, you would not have to starve. In fact, you would find literally dozens of edible natural organisms and plants (fruits, vegetables, roots and seeds) that can be beneficially consumed. Not only can they be consumed, but many are quite delicious.

We will not here consider the numerous marine and fresh water fishes, crustacea and mollusks which are familiar to nearly everyone. Many of the shellfish that are so avidly collected here are also delicious, but of course I don't encourage the consumption of the animals residing in live shells. Also, many can be toxic. The marine aspect will be omitted from this discussion and we will concern ourselves primarily with plants that can be found in the wild parts of Sanibel and are truly nutritious and delicious.

Perhaps the most abundant edible plant is our state tree, the Cabbage Palm, *Sabal palmetto.* Here it is a "no-no" to cut and consume, but it must be included in this list because it has been eaten perhaps more widely than most other wild things. The meristem, or bud, high up in the center of the palm, can be removed with a chain saw or, by hand, with the expenditure of great effort. The commercial product known as "Hearts of Palm" is usually of Brazilian origin, and represents another palm species, very like our own *Sabal.*

A good cabbage would be some twenty-four inches long and four inches in diameter, white or ivory-colored, with the texture of delicate cabbage and a flavor like that of wild hickory nuts. It is undesirable to consume this unless you happen to be present when a tree is knocked down by storm or accident, because obviously to take the center heart out of a tree kills it. "Swamp cabbage," as it is sometimes called, can either be eaten raw just as it comes from the tree, fixed as a salad or slaw, or it can be cooked and served as a vegetable, much as you would serve cooked celery.

Of course there's the Coconut, which is everywhere abundant and truly delicious, but rather hard to get out of its shell without the aid of a machete, unless you know the trick. One way to open a coconut is to sit calmly on your doorstep and pound the pointed end on the sidewalk about five hundred times, and then you can tear the husk off with your hands. It's good exercise, too.

There's another variety of Coconut called the Makapuno, grown in the Philippine Islands which, instead of having juice or coconut milk, has a solid custard-like center not unlike the rich ice creams you can buy at Duncan's of Sanibel. The Makapuno is highly prized in the Philippines and since some of our Coconuts are likely to be attacked in the future by a disease called "lethal yellowing" which is rampaging across Florida, it seems to me well worthwhile to attempt the introduction of Makapuno in the hope that it will prove to be immune to "lethal yellowing", thus solving a potential disease problem and at the same time, upgrading the edibility of locally available coconuts. Three varieties of Malayan "Dwarf" Coconuts are being substituted for the "Jamaica Tall" because these three are resistant to this serious disease. All are varieties of one species, *Cocos nucifera,* as is the Makapuno.

Somewhat like aspargus is the curled meristem or "fiddlehead" of the large Leather Fern. This is the two or three-inch wide fiddle-shaped growing tip coming out of this great fern's center — destined to be a new frond unless you pluck it and eat it first. Meristems can be snapped off at about six-inch lengths and kept crisp in a plastic bag stored in the refrigerator until needed. Cook these in a

Photos, Mark Twombly, *Island Reporter*

Carol Kranichfeld, left, sampling a wild morsel. Susan Brooks, right, enjoying a bountiful vegetarian repast — all ingredients of which are found on Sanibel. These college students were assigned to the author by the Conservation Foundation to help research "Sanibel's Bounty." Both received college credit for their efforts.

shallow frying pan with a little salted water, butter and lemon. This makes an excellent delicately-flavored vegetable, and the price is right.

Another good one is Sea Purslane, which is found all over the place in our wetlands and on the beach ridges. This thick-leafed, juicy plant can be consumed raw, as a little snack while hiking, but it's good when cooked also. It's rather salty, so those on low salt diets should eat it only when cooked and then only after changing the water several times.

There is a little Japanese Lantern-like member of the tomato family called the Ground Cherry, *Physalis.* It grows throughout the disturbed parts of our filled wetlands areas. You will find this little fruit very tasty, if you get there ahead of the worms. You simply break open the yellow "lantern" and find a tiny round fruit that looks like a small yellow tomato.

Everywhere on Sanibel are Spanish Bayonets, a kind of *Yucca.* These, in the spring and summer — some early, some late — present tall white inflorescences having white or ivory-colored thick petals which are really great in salads. They have a nutty flavor, rather like fresh-roasted cashews. *Yucca* petals can be gathered, wrapped in a damp towel, chilled in the refrigerator and used to garnish salads. They are also excellent when used in place of crackers or chips with dips. Of course you must be careful to eliminate the ants which are rather plentiful in these inflorescences, unless you wish to have an added protein supplement with a formic acid flavor.

A salad can be further garnished with the beautiful *Hibiscus* flowers which add a great deal of color and can be used entire, as decorations, or snipped with scissors into strips and tossed throughout. These petals add considerable nutrition and a delicate peppery flavor which is really desirable. Although the *Hibiscus* makes a fine snack, it's perhaps undesirable to consume your hostess's centerpiece should dinner be delayed. I once had an experience of this kind and it was not a happy one.

In the various xeric or desert-like regions of our Islands can be found quite a number of species of cactus. The cacti that I have found on Sanibel all produce delicious and edible fruit. Care must be taken in removing the spines from cactus fruit, but after this is accomplished, all that remains is pure enjoyment.

The great Night Blooming Cereus which blooms so abundantly here in the summertime, with ten-inch long flowers as big as dinner plates, produces a delectable fruit with a consistency rather like watermelon, but with the fragrance and flavor of a fine French perfume.

The so-called Rope Cactus produces an equally delicious edible fruit. The great Peruvian Cereus also has a delicate watermelon-like textured fruit and of course all the various kinds of prickly pears produce edible fruits. These purple colored fruits can be seen at any season in our xeric wild lands, but great care must be taken, because the little patches of spines contain hundreds of hair-like prickles. These can get into your skin and you will live to regret your rash handling of this fruit. A sharp knife and deft hand can easily remove the spines. Then both the outer fruit and the inner pulp containing seeds can be eaten. The taste here is like that of pomegranate flavored with cranberry.

Before we leave the prickly pears or *Opuntia*, we must mention that the pads are also edible. One variety commonly grown here, called the "Burbank Spineless", which is alleged to have been developed by Luther Burbank but probably was not, is used as fodder for cattle. It can be eaten with gusto, raw, right from the plant. Select very young pea-green pads. They have a delicate and good flavor, but the slight sliminess might be distasteful to some. It offers no problem to those of us who grew up in the South and were early introduced to okra. But any distastefulness can be overcome when you again consider that the price is right.

The author with an ***Agave,*** *the species from which the Mexicans distill that fine tequila.*

Photo, Mark Twombly, *Island Reporter*

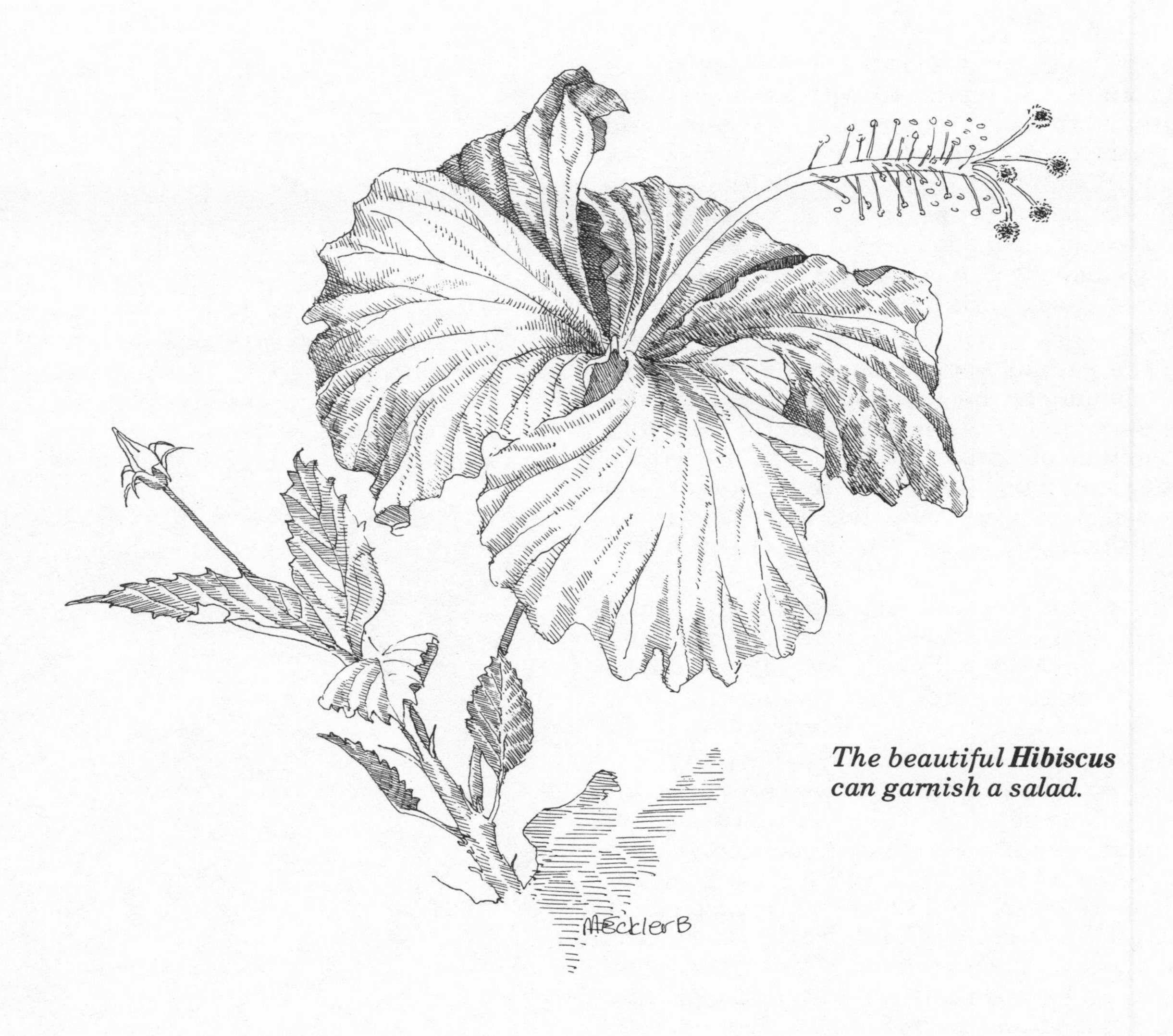

The beautiful ***Hibiscus*** *can garnish a salad.*

Pads of the Burbank Spineless — or for that matter, any of the species of Prickly Pear — can be sliced and cooked rather as you would cook French-cut green beans. As a matter of fact, that's what they look like when they're prepared. Sliced cactus pads can also be fried in an egg-breadcrumb batter and this is good too. The taste and texture are reminiscent of fried eggplant.

The Coconut tree is useful in another department also. The inflorescence can be cut transversely and the sap collected in a jar and later fermented and made into a delicious beer called coconut toddy.

There is another plant that can help out at cocktail time and that's the *Agave* or Century Plant, of which there are a half dozen kinds here. It is from plants of this genus that the Mexicans create that delectable nectar known as tequila. Mescal, slightly different, is also made from *Agaves.* In this country we're not supposed to distill our own booze, but there's no law against making pulque, a fermented beer also of *Agave* origin.

I have been lucky in my gastronomical explorations. Only a few stomach aches and never a really serious reaction. I do have the wit not to experiment with the strange but beautiful mushrooms that we see here.

I certainly recommend my hobby to anyone interested in free eating, a pastime which gains more significance as each day passes in our inflationary economy. I haven't begun to mention everything that's edible here — for example, there's a little cucumber. Also we find tomatoes in the wild; various grasses are good. There are a lot of cattails and they have several edible parts. There are a number of marine algae that are edible. Some bromeliad inflorescences besides the well-known pineapple have delicious flavors, but not much food value due to their small size. So the list goes on and on.

Let me conclude with the final note that it is my intention to look into seeds as possible grain substitutes. Our so-called Australian Pine *(Casuarina)* seed is here literally in tons, and I tried it on my cereal and found that it tastes and looks a bit like wheat germ. I haven't yet found anybody who can analyse it and tell me whether it is nutritious or not, but a recent biologist visitor from Queensland told me that many parrot-like birds of Australia feed extensively on *Casuarina.*

Also, as I walk through the wild parts of Sanibel and pick up googols of Sand Spurs, I'm reminded of the method used to prepare hominy (or hominy grits which are better known in these parts) which is made from corn. I believe the outer part of the corn is leached away with lye. Possibly we could come up with a use for the Sand Spur by leaching away the spines and thus discovering a use for this otherwise horrible pest whose only apparent function is to serve as a nostalgic reminder to the visitor that he has been to Sanibel.

The plants mentioned here, except the Sand Spur and Australian Pine seed, are all accepted edibles. Experimentation beyond that list may be undertaken by botanists or other venturesome experts, but certainly not on my recommendation. Those who take up this hobby should be prepared to undertake extensive reading in the botanical literature.

18 *THERE ARE POISONOUS PLANTS TOO*

Sanibel can be called one of the great Poison Ivy capitols of the world. One of the favorite habitats of *Toxicodendron radicans* is the trunk of the Cabbage Palmetto or Sabal Palm. Nearly every such trunk in the wild has its supply of Poison Ivy, and most of those growing in residents' yards either now have, or have had in the past, a plentiful supply climbing the trunks.

Poison Ivy is quite a beautiful vine, with an odd three-part pinnate leaf -- a shiny, beautiful green. But it does turn bright red about Christmas time. In the past, certain County politicians whose popularity had waned learned ed to beware of beautiful little red-leafed potted plants sent at Christmastime by disgruntled ex-constituents.

And one visiting lady of my acquaintance had quite a bit of Poison Ivy in a colorful bouquet of wild plants which she had lovingly gathered one winter morning on Sanibel.

Dr. Jean Gentry points out that Sanibel's Poison Ivy is somewhat different in appearance from the plant as it is usually seen "up north" and varies even from the illustrations usually given in books. So it is wise to learn to identify this plant correctly.

It can cause a great deal of unpleasantness through direct contact, through contact with objects that have been contaminated, or even through smoke from a burning trash heap.

Another plant to consider is the Mango, *Mangifera indica.* It comes to us from India and each summer produces a delicious and very popular fruit. But there are those among the general population who are sensitive to the fruit and the sap from the tree, which can cause swelling and itching. The flowers produce an irritating substance which can cause serious respiratory reactions. The burning of Mango trash can also cause problems. Yet the fruit is absolutely delicious and much-loved by millions of people throughout the tropical world. Queen Victoria said it was the finest fruit of her realm.

Many people in whom the Mango causes allergic reactions find that they can eat the fruit if someone else peels it, since the most allergenic portion of the fruit is associated with the tough skin and the exudate at the fruit stem.

Of course, some people are allergic to the fruit itself, and should exercise caution in this regard. It should also be added that some people can eat and enjoy Mangos for years and then suddenly become allergic to them.

Next is the Brazilian Pepper, *Schinus terebinthifolius.* This plant, as well as the ones previously mentioned, Mango and Poison Ivy, all belong to the family Anacardiacae. Anacards of other kinds are also often skin or respiratory irritants, but these three are the principal ones found on Sanibel.

The Brazilian Pepper may be one of the most abundant trees on the Island, and is listed by Richard Workman as one of the "Big Three" weed trees. It is doing untold damage to the environment, besides creating health problems for many Sanibel Citizens. The conservation effort on Sanibel is likely to be lost by 1984 and, if so, it will be largely due to this tremendously destructive species.

One possible hope lies in the study of this species in its native country to seek some specific biological control. Having spent many years as a naturalist in Brazil, I would like to return there to conduct such a study and seek out natural controls.

For some, the resin of the Brazilian Pepper produces a serious dermatitis, which might include eye problems and swelling of the hands and face. Brazilian Pepper plants can be removed rather easily from a property simply by cutting them down and repeatedly painting the stumps with stump-killer, or, better yet, drilling large holes in the stumps and filling them up with strong lye, kerosene or crankcase oil.

But here on Sanibel we will probably never be able to get away from this tree completely. When it blooms, it produces something which is wafted on the air. Your neighbor's tree, down the road, can cause you respiratory distress even though you may have thoroughly cleaned up your own property.

The Brazilian Pepper is not a true pepper, but there are some unscrupulous pepper merchants in the world who use the ground dried fruits of this species to dilute black pepper.

A popular exotic landscape plant here is the Oleander, *Nerium oleander.* This produces a digitoxin-like drug which, in the body, produces dizziness, lowered pulse, cramps, vomiting, pupil dilation and — if you consume enough of it — eventually paralysis, unconsciousness and death. The problem can arise from chewing the leaves or flowers, from inhaling smoke from the burning of Oleander leaves and prunings, from eating poisonous honey produced by honey bees, or from roasting marshmallows on a stick from an Oleander bush.

The Oleander has long, lanceolate leaves and pink, red or white flowers with many variations. It is a familiar sight on Sanibel and is frequently grown as a hedge. There are some old plants along Wulfert Road, left over from years ago when this area was farmed.

Oleander flowers smell like cheap dime-store perfume. The plants themselves, if not given a lot of care, will show much dead wood. All in all, it is my opinion that this plant is so undesirable that it would be well if it were eliminated from our Island.

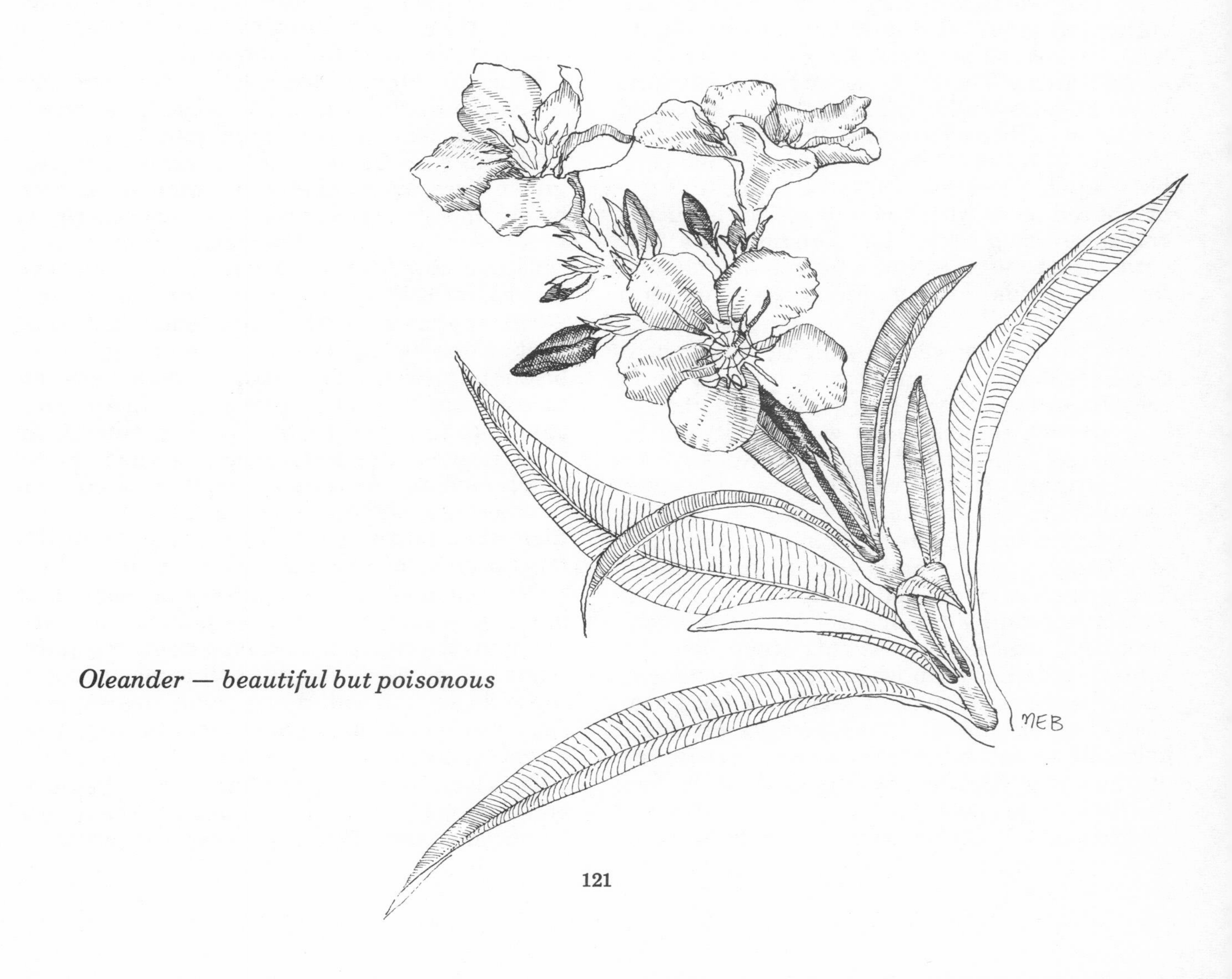

Oleander — beautiful but poisonous

Pokeweed, *Phytolacca americana,* is a familiar plant which grows quite large, having a thick red stem and foot-long lanceolate leaves. It is sometimes called inkberry because of its ripe fruit which is dark purple, nearly black. This plant, especially in its root, contains toxic alkaloids and will, when consumed, cause cramps, purging, possible convulsions and paralysis. Some deaths have been recorded.

Lantana occurs here in several species, some as native bushes, others as colorful garden varieties. Its unripe bluish-purple fruits have caused death, and children should be taught to recognize and avoid *Lantana.*

One of the most attractive exotic vines that has crept into Sanibel and, for that matter, almost all of South Florida, is the Rosary Pea, *Abrus precatorius.* This feathery little legume is quite lovely and can be found climbing Cabbage Palms, Mango trees, abandoned citrus trees, telephone poles in vacant lots and, in fact, many sorts of bushes in disturbed or once-cultivated areas. It has a very small pinnate leaf — only about four inches long — with many leaflets. The fruits, as one would expect, look like peas, and these are in a little pea pod about an inch and a half long. Pods occur in clusters of up to 20 in one ball and when ripe, burst open to expose the quite beautiful little bright red seeds with black eyes. This beautiful and interesting little plant provided the first carat weight used in India by gem merchants, for the seeds are of remarkably uniform weight.

The seeds, however, are highly toxic, containing substances known as abrin and abric acid. If one were to swallow a seed, it might not cause a problem because it would probably be eliminated entire, for it has a hard protective outer surface. But if a child were to chew a seed, it could readily cause death.

When I was a boy in Florida, it was popular for little girls to string beads of Rosary Peas. The pea was harvested before it was completely ripe and was not yet red, but a bright pink with a dark spot. At this stage they are somewhat larger than when ripe. A needle can easily be thrust through the not yet hard pea. The children made necklaces and bracelets using these and other colored seeds. Later, as drying occurred, the peas would achieve their full brilliant red color.

Many small children found this to be a pleasant pastime, but according to Julia Morton, botanist of the University of Miami, the introduction of the poison into the system by way of a pricked finger is many times more toxic than by ingestion through the G.I. tract. Therefore, stringing of these "beads" is a practice that should not be permitted. In fact, it is my understanding that Florida law now prohibits the use of these peas as jewelry. The practice still goes on in the West Indies and other parts of the tropical world. But Miami customs inspectors are now trained to recognize the Rosary Pea, and such jewelry is confiscated when found.

The Coral Plant, *Jatropha multifida,* is grown all over Sanibel. It has a palmate ten- or eleven-lobed leaf, each leaflet being pointed and lanceolate, reminiscent of marijuana. The flower stalk looks rather like the red Mediterranean coral used for jewelry, and the fruits are in capsules containing three highly poisonous seeds which, unfortunately, taste good. The poison is called jatrophin. Children eating these seeds become very seriously ill and may need hospitalization.

Other *Jatrophas,* the Physic Nut and the Bellyache Bush, seem to cause the same severe toxic reactions that the Coral Plant does.

The Castor Bean, *Ricinus communis,* was grown here on Sanibel as a commercial crop before and during World War I, when castor oil was needed as an instrument lubricant during that conflict. Today on our Island we have many Castor Plants left over from those days.

This palmate eight-or-nine-lobed leaf with fruit borne in clusters of burr-like green or bright red spikes will split open and interesting tick-like seeds will be thrown out. These seeds are mottled, light or brown, and sometimes their resemblance to a large engorged female tick is striking. As a weed tree, it is common on Sanibel, but there are also a number of horticultural forms that have very beautiful flowers and fruits that are widely used.

Castor Oil is not, of course, poisonous and has been used in medicine for many years. But the seeds contain a toxin called ricin. When the seeds are pressed, the pure oil comes out and the ricin is retained in the cake. Ingestion of ricin can cause anaphylactic shock and, conceivably, death.

The Coral Plant, Physic Nut, Bellyache Bush and Castor Bean are all members of the family Euphorbiaceae. There are three other Euphorbs that grow on Sanibel as introduced ornamentals from Africa. They are the Candelabra "Cactus", *Euphorbia lactae;* the Pencil Tree, also called by its Masai name Manyara, *Euphorbia tirucalli;* and the Poinsettia, *Euphorbia pulcherrima.* In these three plants, it is the white sap that causes severe eye problems and temporary blindness and may also blister the skin.

The familiar Poinsettia is used very widely at Christmas time, but is a plant that should be handled with caution because it is truly poisonous. We also have two beautiful little wild Poinsettias on the Island and I believe their sap is also poisonous.

You can always tell a true cactus from a *Euphorbia* by the white sap of the latter. A pinprick in the plant will tell you whether it is poisonous or not, for the white sap appears immediately.

This is not to imply that all plants having white sap are poisonous; many are not. Some of our exotic shrubs having white sap were introduced into Lee County by Thomas Edison, in his search for an easily-grown source of rubber.

A beautiful little wild plant, really a wild cucumber and often called Balsam Pear, *Momordica charantia,* is found here and is often attractive to children who may be tempted to pick it and eat it. The seeds and the orange body are poisonous and do cause internal problems, yet many people suck the seeds, the red part (the aril) being edible. This little fruit grows in the wild parts of Sanibel.

It shouldn't be necessary to caution people about mushrooms, but it is! There are among us mycologists who study mushrooms and know which ones are harmlessly delicious and which ones will kill you. If you are not a qualified mycologist, it is recommended that you buy your mushrooms at Bailey's Store. Even the experts make mistakes. In 1974 in Michigan, two qualified mycologists lost their lives because they made a wrong identification. Dr. Jean Gentry reports cases of Florida State University students who were positive they knew the difference, then poisoned themselves with toxic mushrooms.

Most poisonous mushrooms cause an intoxication like that caused by atropine, which can be rapidly fatal if not treated promptly.

Several plants in the family Solanaceae, the same family that includes the delicious tomato, are poisonous. The Chalice Vine, *Solandra nitida,* contains alkaloids that are powerfully narcotic but this great golden cupped flower can easily be identified.

Some Jasmines, such as *Cestrum,* are bad. It was *Cestrum diurnum,* the Day Jasmine, that caused this chapter to be written. A child ate some, causing consternation but, fortunately, no severe medical problems, and I was urged to write on this subject. The berries, if eaten in quantity, can cause a mild atropine-like reaction.

The Devil's Trumpet, a *Datura,* a large pink-flowered plant, and the so-called Angel's Trumpet, also a *Datura,* with pendulous large six-inch long flowers, are dangerous plants. They are grown to some extent in gardens here, and children should be cautioned not to touch them.

The Cajeput, *Melaleuca quinquenervia,* is a plant that gives off respiratory irritants during the blooming season and is one of the previously-mentioned "Big Three" weed trees listed by Workman.

The familiar Oyster Plant, *Rhoeo spathacea,* causes rash and sometimes respiratory difficulties. The eyes may smart and become inflamed after one carelessly handles the plant which nearly every garden on Sanibel, including mine, is cursed with. Similar medical problems are caused by the Purple Queen, *Setcreasea purpurea,* the familiar herbaceous purple ground cover plant that is grown so widely here. Neither of these members of the Commelinaceae presents a very serious threat, but be warned not to be careless in handling them.

Another family which is widely-grown here is the Araliaceae, the so-called Aralias. There are many species, mostly in the genus *Polyscias.* They cause a rash and throat and mouth irritation when handled or chewed.

The so-called Century Plants exist on Sanibel in several species, all of the genus *Agave,* belonging to the Amaryllis family. These large succulent plants, which send up huge inflorescences that sometimes reach 35 feet, can be harmful to children as the juice causes burning, rashes and blisters. If you get the juice in your eyes, you can experience serious irritation and even temporary blindness.

Of course the main danger of the *Agave* relates to the terminal thorn. These spikes or thorns occur on the ends of most Agave leaves; some of them are also armed with barbs along the edges of the leaf. These present a danger to children and pets, and if you culture this beautiful group of plants for your garden, it's sensible to take a pair of nippers and cut off the terminal thorns.

Many Aroids, that is members of the family Araceae, such as the Elephant Ear, *Xanthosoma,* the *Alocasias,* some *Anthuriums,* and the well-known Dumb Cane, *Dieffenbachia,* are also poisonous. The *Dieffenbachia* gets its common name, Dumb Cane, from the fact that if eaten, it causes a swelling in the throat, rendering one incapable of speech. This plant and other related ones are grown very widely in our horticulture; many homes on Sanibel have them outside as well as potted for house plants. It's well to teach one's children not to touch. This plant can also be fatal to pets who chew it. Its Brazilian name, *Comigo Ninguem Pode,* roughly translates to "Don't fool with me," for in its native country, its dangers are well known.

Another Aroid is the great so-called *Pothos aureus,* now called *Rhapidophora aurea.* This is the yellow, green and white-leafed vine grown very commonly as a houseplant which, when placed on a coconut tree, will attain huge size with great split leaves up to a foot or two in length. It will climb and cover the coconut tree in a very beautiful way that is popular in horticulture.

This form is closely related to the many *Philodendron* species, grown both outdoors as self-heading varieties and indoors as houseplants. Children and pets tend to chew some of these plants and can get a very serious irritation of the mouth and lips. *Philodendron selloum* has been identified as a plant that causes severe irritation in children who chew its leaves. This writer takes credit for having introduced this species to the United States from Brazil many years ago, so any problems with *Philodendron selloum* can be blamed on me.

Erythrina is a genus of leguminous mostly exotic trees that are grown here sometimes. One is a native — *E. herbacea* — a beautiful tree called the Cherokee Bean. It is one of those valuable plants listed for protection in the Sanibel Comprehensive Land Use Plan, a landmark document that will go a long way toward environmental protection here, if all goes well.

The seeds of all *Erythrina* species are red and dangerous and children should be cautioned not to eat them. They contain alkaloids which are used as rat poison in some places. One of the species of *Erythrina* is called Immortelle because it is durable and makes good coffins.

The trees and plants listed and discussed here are considered by me to be the ones most likely to be encountered by children on this Island, but I want to emphasize again that there are probably a hundred other species which could cause trouble. A knowledge of local botany is important protection and will also prove interesting to parents willing to make the effort.

Botanical familiarization should be an important aspect of the education of any child growing up on Sanibel Island.

19 *THE HORSE-FOOT (ALIAS THE HORSESHOE "CRAB")* Miracle Worker of the Sea

The Horse-Foot is an ancient animal which has come down through the ages almost unchanged for over 200 million years. It has no living relatives in the New World, and only two other genera exist anywhere in the rest of the world — both in Asia. The Horse-Foot is commonly called Horseshoe Crab but it is not in any way related to the true crabs which are crustaceans. Nor is it properly called "Horseshoe" for this is a corruption of "Horse-Foot" which is what early settlers in America called it. Actually Horse-Foot is a more realistic handle, for the arched carapace covering the cephalothorax is indeed somewhat the shape of a horse's hoof. The animal bears no resemblance at all to the familiar U-shaped iron horseshoe.

This strange animal belongs to an almost extinct class called Merostomata, within which the only survivors are in the order Xiphosura, which simply means "swordtail". It is interesting to note that its closest living relatives are the arachnids, the spiders and scorpions.

Horse-Foots do not occur very widely in modern oceans. Our own species ranges from the Bay of Fundy to Key West on the Atlantic Coast and in the Gulf of Mexico in such places as Sanibel and in other scattered localities around as far as Yucatan in Mexico. It is technically called *Limulus polyphemus,* a name that refers to the beast's interesting eyes — two large compound eyes on each side (*Limus* means sidewise glance), and two single eyes in the middle part of the dome (Polyphemus was the Greek god with one eye in the middle of the forehead). The larval stage even has a third pair of eyes! These are on the underside and cease functioning when the animal matures.

The two allied genera called *Tachypleus* and-*Carcinoscorpius* (an unpleasant name if I ever heard one) are found from Korea and Japan along the China coast south to the East Indies and as far west as India. *Carcinoscorpius* is more of a fresh water animal than our own Horse-Foot in that it travels far upstream in some Asian rivers such as India's sacred Ganges system and Kipling's "Road to Mandalay," the Irrawaddy.

In some seasons of the year *Limulus* is very abundant around Sanibel. You may have seen great heaps of their cast empty shells on the beach. Other times, due to Raccoon predation, many dead animals are to be seen along the Darling Memorial Drive.

Perhaps the one thing that best demonstrates the Horse-Foot's relationship to arachnids is the fact that, in common with a lot of spiders, it does not shed its old shell through a mid-dorsal opening. Instead, a Horse-Foot splits along the forward rim of its carapace and sort of crawls out of this broad but vertically shallow slit. The cast shell looks very lifelike and is often so completely intact that many people confuse these perfect specimens with dead animals, which of course they are not.

Horse-Foots are absolutely harmless except for one thing — they do have a tail appendage called the telson that is hard, sharp and barbed. This is attached to the posterior end of the abdominal region which contains the gill books which are used for oxygenating the blood and also can be flapped rhythmically to propel the animal when swimming. The caudal spike may be in a vertical position while the animal lies half buried in the mud or sand in shallow water, and bathers have, on rare occasions, received painful wounds from stepping on them. However, the animal does not use the spike to attack, so Horse-Foots are quite harmless in spite of their rather formidable looking anatomy.

In early times some coastal Indians used these spikes as spear points. When fastened to long straight shafts, they were said to be quite effective in gathering fish and crabs.

The eggs and meat of the Horse-Foot, what little there is, were also consumed by Indians. On Martha's Vineyard, Indians used the pinchers of the feet in their jewelry.

American settlers used the Horse-Foot to feed their stock, principally swine and chickens. It was also dried and ground for use as fertilizer.

If you tip over a Horse-Foot you will see a number of dangerous-looking appendages which are all really quite harmless. The foremost pair are chelicerae; then come the two pedipalps and finally the four pairs of legs. The legs are equipped with pincers at their "ends" which are used for grasping and breaking down food and passing particles along toward the mouth where they are ingested.

It is rather easy to tell the smaller male *Limulus* from the larger females because the pedipalps of the male have enlarged hooks especially adapted to grasp the trailing edge of the female. Sometimes a regular "conch train" of Horse-Foots may be seen attached to one large female, following her around and being partially propelled by her, awaiting the time when her eggs are ripe and she goes out onto the beach on a high tide to burrow and lay in the sand or mud. A large female may be 30 inches in diameter, although such huge examples are not found in this part of their range. Males are smaller with perhaps a 12-inch diameter at the most.

In some Asian cultures the green unlaid eggs of *Carcinoscorpius* are considered a great delicacy. As in the case of the "Fugu" or puffer fish of Japan, eating Horse-Foot eggs or "Mimi" is like playing gastronomical Russian roulette. The not-infrequent Mimi poisoning causes cardiac palpitation, abdominal pain, generalized weakness, paralysis and sometimes loss of consciousness. Mortality frequently results and usually when that happens it is less than a day later. Yet this food is widely consumed. Specially licensed and trained Mimi chefs handle the selection of eggs and their artful preparation.

Horse-Foots can crawl along the ocean bottom and are often found in deeper water out in the sea. They can also swim belly up along the surface of the water, moving legs and gill books for propulsion. If they fall to the bottom on their backs, they may right themselves by flipping the mobile telsons.

Charles LeBuff has found the Horse-Foot to be a frequent item in the diet of the Loggerhead Turtle. "I have opened several dozen Loggerheads that have been washed up dead on Sanibel beaches and have examined the contents of their digestive tracts and have found that the Horseshoe Crab is a major component of the diet of this threatened sea turtle," he says.

Could it be that the turtle eats them only when desperate for food and is sometimes killed — a la Mimi poisoning?

Sanibel's infamous red tide, in addition to "doing in" hundreds of thousands of other animals, occasionally destroys many of these interesting "living fossils." On such occasions, when the beach may be littered with dead ones, it is rather unsafe to walk barefoot because one can receive a nasty spike wound.

Recently the Horse-Foot has proved to be of great interest to science. In these days of energy shortages and future anticipated crises, the Horse-Foot's compound eyes, located on either side of the foreward dome of the carapace, have created a great deal of interest among some physicists. Each compound eye is made up of some ten or fifteen individual "eyes" or groups of retinal cells surrounding a rhabdom, similar to the eye of the insect. The resulting "dome" of "facets" is the most perfect light gathering design to be found in nature. This property can be easily demonstrated by holding a shed shell in the bright sunlight. The eyes, from the inside, show up as bright "pearls" in the otherwise opaque dome of the carapace.

Design research in this area is going forward at the University of Chicago and elsewhere, where this ancient creature may yet contribute to concepts and patterns for devices that gather solar energy, an energy source which inevitably must prove to be of eventual great importance to man in the not so distant future when our fossil fuel is no more.

The large size of the cone cells of this great compound eye has been found useful by ophthalmological researchers who are studying the very nature and basics of optical vision. These cells are so large that one can be isolated electronically; i.e., an electrode can be attached to it and the actual conversion of light energy into electrical (nerve) impulses can be observed and measured by employing sophisticated electronic equipment — all connected to a single

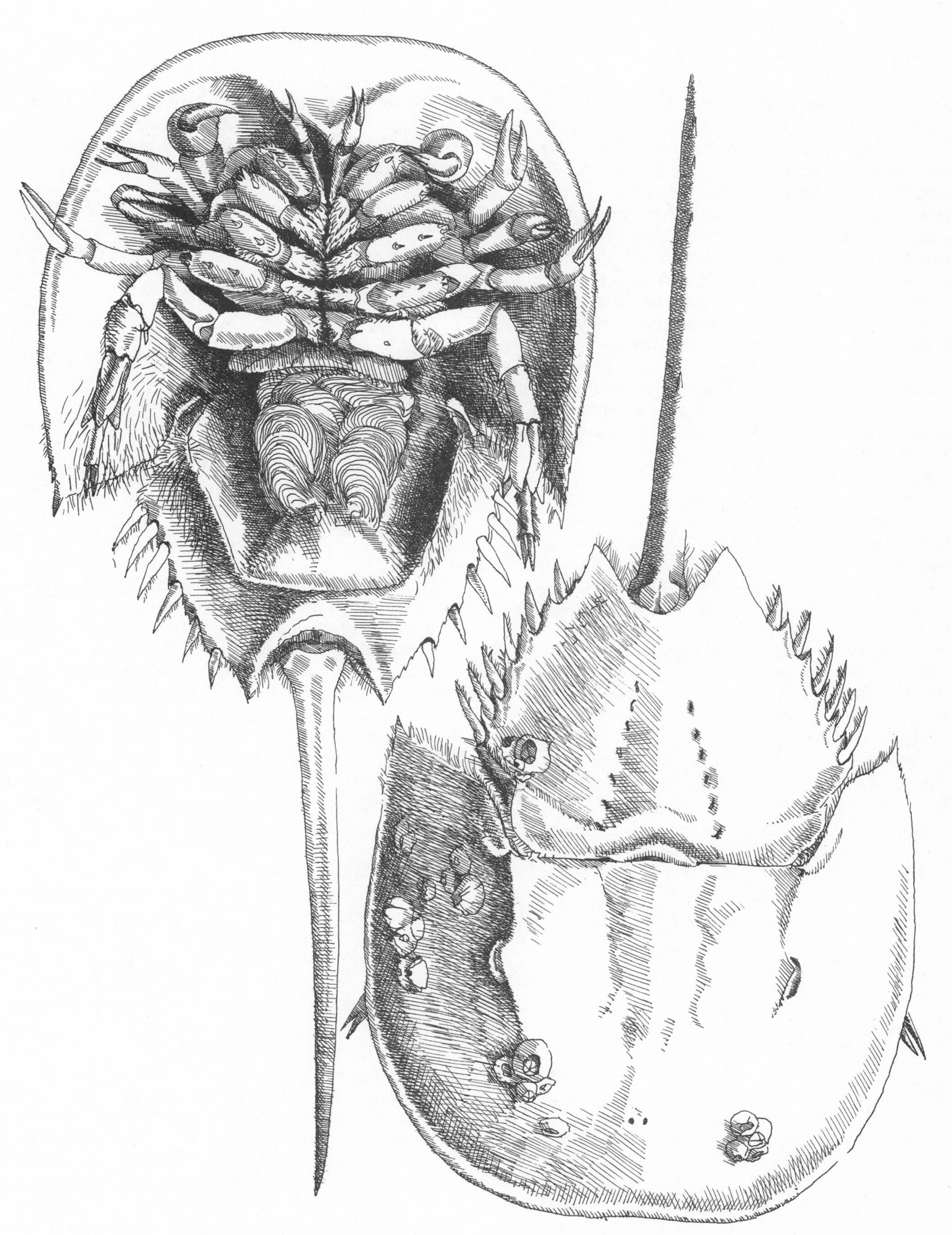

Horse-Foot, miracle worker of the sea. This is a male. Note the second pair of appendages. They are modified to enable the male to clasp the carapace of the female.

such cell. The possible benefits from this line of research are great.

Another Horse-Foot marvel relates to the creature's unique pale blue blood. Men and dogs, for instance — and other mammals — have red blood pigment that is iron-based (hemoglobin). Horse-Foots have a blood pigment that does not contain iron, but copper instead, and this hemocyanin, as the blue pigment is called, has proved to be of immense interest. It is used as a stimulator of antibody production in experimental animals and man. Furthermore, the Horse-Foot's only circulating blood cell is used in the detection of endotoxins, those poisons that are produced by gram-negative bacteria and which often contaminate parenteral or injectable pharmaceutical products. The extract or lysate of *Limulus* blood cells (amoebocytes) is now being widely used in the detection of such contaminants in the pharmaceutical industry and other medical areas. Certain Horse-Foot blood proteins can be used in laboratory tests to distinguish between normal and leukemic human white blood cells.

Environmental pollution can also be detected through the same substance, for bacterially-produced endotoxins in the environment often denote pollution and this test can point to the need for corrective measures.

One chemical company produces a commercial product of highly purified material from the *Limulus,* that sells for a hundred dollars a milligram. Consider that the ink needed to print the word "milligram" weighs about a milligram and you get a suggestion of the values involved.

Another fascinating fact is that the *Limulus* hemocyanin is the world's largest molecule, having a molecular weight of eight and a half million! The common hydrogen molecule has a weight of two, so you can see that the contrast is tremendous. If a hydrogen molecule were represented by a foot-long Garter Snake, the comparative size of the hemocyanin molecule would be that of a huge Python that could stretch from Sanibel to Buffalo, NY, where lives Dr. Elias Cohen, a biologist who has done a lot of research on the Horse-Foot blood proteins and who has shared his knowledge with me. Cohen often comes to Sanibel and I have had the great pleasure of being on a number of field trips with this informed and stimulating scientist.

For many years on our coasts, the Horse-Foot has been cordially hated for destroying some shellfish and it was even on bounty in Massachusetts. Now the tide is turning; the animal's value is being appreciated, and among enlightened people there is great concern for its conservation. So when you see one on our Sanibel beaches — or in the mangroves of the bayside — don't disturb it because it is proving to be one of the most valuable sea creatures in existence.

20 *THOU SHELL NOT* The Shell Game That Everybody Loses

One wildlife community of our Island's marine environment that is really suffering under human pressure consists of the many seashell species which adorn our beaches and the sea around us, shells that are among Sanibel's most treasured resources.

Recently there has been much controversy concerning what I believe to be an intolerable, unconscionable activity — the wanton taking of live mollusks. This, the most recent of a long series of controversies on the subject, was triggered at the 1976 Shell Fair, when I was attracted to an unusual specimen being offered at a table heaped high with shells, all of which had been taken alive. Upon asking what this different-looking shell was, the immediate reply shot back, "Fifty cents." From this commercial retort it became obvious that some of the values concerning Sanibel's natural resources are becoming distorted.

When I first came to know Sanibel Island, some 40 years ago, it was a very much different "Seashell Island" than it is today. For one thing, there were many, many more shells than there are today. For another, it was possible to find much larger specimens of nearly all species than are encountered these days. For example, the Florida Crown Conch or King's Crown, *Melongena corona,* which is still plentiful in Sanibel waters, used to occur in larger numbers here and individuals attained much greater size.

I know of one specimen, seven and a half inches long, which was collected many decades ago. I haven't seen any approaching that size in recent years. Why? Probably they don't have a chance to reach their maximum size because of the large numbers collected. Yet the collection of this species in great numbers — even boatloads of them — goes on today and is defended by those who would reap a harvest from the sea for profit.

It is inappropriate for the human creature to consider himself civilized if he is willing to sanction the senseless killing of wild animals for fun or money.

Defenders of recreational and commercial slaughter of wild things for profit aggressively put forth false and distorted arguments based upon misunderstandings, often intentional, of the principles of "the balance of nature," "survival of the fittest," "conservation-mindedness" and, of course, "free enterprise."

There is a publication of the State of Florida entitled *Environmentally Endangered Lands Plan.* This book has a section dealing with endangered, threatened or otherwise diminishing species of Florida animals and plants. It lists seven pages of species, each page containing from fifty to one hundred kinds of plants and animals, as being in these categories.

The International Union for the Conservation of Nature offers loose-leaf notebook services in the area of endangered and threatened animal and plant species throughout the world. There are literally thousands of forms in these categories. Most of them have reached their state of crisis through man's agency; often to satisfy a wide range of selfish needs which one must presume cannot adequately be satiated without bloodshed or other destruction.

Some who do not actively indulge in such behavior still exhibit, through tragic disinterest and insensitivity, a tolerance for these acts. In a truly civilized society, killing for profit or fun would not be permitted.

The unnecessary senseless killing of living creatures is simply not an act of healthy fun. It is certainly not prestigious, but rather it is primitive. I do not refer to the humane slaughter of domestic animals raised for man's nutrition. To my mind, there is a great difference between wild animal destruction and the husbandry and exploitation of man's domestic commensals.

I was impressed by an article in the January, 1976, issue of *Audubon* by Richard Ellis, a well-known artist whose works can be seen in Sanibel's own Schoolhouse Gallery.

The title of Ellis' article is "Why I Became an Ex-Shell Painter." His thrust is an indictment of overuse of a valuable species, in this case the Queen Conch, *Strombus gigas*.

He takes issue with the non-scientists, the shell collectors, the conchologists — those who collect cabinets full of shells as they would shelves full of fine china — but are often unconcerned with the shells' biology and unaware that for each "item" in the collection, a living creature died.

Such collectors really seek only those exciting, unusual examples, the trophy shells, which are, of course, the very ones which should be left in the gene pool, for they carry the best genetic potential for the maintenance of quality in future generations.

My son, David Campbell, who heads the Bahamas National Trust, the leading conservation organization in the Bahama Islands, tells me that *Strombus gigas* is becoming scarce in some parts of the Bahamas and Turks and Caicos Islands, due to over-collection. It is now contrary to Bahamian law to export these shells, or for that matter, any specimen of marine life without prior permission. *Strombus* has been consumed as chowder, salad and even steak throughout the Bahama Islands for generations — ever since large numbers of Tories left New York to colonize those islands two hundred years ago.

Today it is becoming more difficult to find conch in the Bahamas, although skillful native fishermen still manage a harvest from the remnant population. But there is real concern in the Bahamas for the future of this once abun-

Drawing, Ann Winterbotham

The official Sanibel City Seal is like this except that the shell pictured is a Fighting Conch. Some feel that the Fighting Conch does not properly represent all Sanibel Citizens. Hence, in this unofficial version, ***Mercenaria,*** *the "Symbol of Greed," is shown in place of* ***Strombus alatus.***

dant and very valuable species.

Other places also completely protect their sea shells. In the Galapagos Islands, which were much-brutalized during the last century when whalers and pirates rendered the great land tortoises almost extinct, there are now very stringent laws imposed by the government of Ecuador, which prohibit the taking of *anything*, dead or alive, from any beach of the Galapagos. The Ecuadoreans, supported by enlightened conservationists of Europe and America, are so concerned about the preservation of the islands' natural values that when one travels from island to island, even the cuffs of trousers are examined. Shells, seeds, sand and soil are removed so that there can be no contamination of one island by material from another. The practice also insures that there will be no removal of any natural product from any of the islands.

The Hawaiian Islands, hardly less environmentally damaged than were the Galapagos Islands, now have strict regulations of a like nature on some of their marine preserves. And California now prohibits live shelling without appropriate permits.

R. Tucker Abbott, internationally known authority on malacology, is the author of an excellent shell guide, *Sea Shells of North America.* In his section on collecting mollusks, Dr. Abbott suggests that with the huge increase in the number of shell collectors it is well to consider the conservation of shells. He recommends that habitats not be disturbed, that cover be replaced after collecting, and that one always leave some specimens. He suggests that if one has a successful collecting trip, it may be due to the consideration of the preceding collector.[1]

Richard Workman recently encountered a sheller on the beach carrying buckets of freshly-collected live shells. When Workman confronted the man and explained that such exploitation was depleting the natural resources of the island, the collector replied, "Yes, I know; that's why I want to get mine before they are all gone."

It has been argued by some that live shells washed up on the beach have only a brief life span left, having been discarded by the sea. This is not generally true. It is well known that beached shells can survive. There are many records of live shells taken from heaps on the beach which have actually reproduced in an aquarium. Many Island aquarists have had this experience. In other words, because a live shell has been cast upon the beach by the sea is no reason to assume that it is doomed. If replaced in a proper marine habitat, it may well survive. Martha Reynolds, a skilled marine aquarist and conservationist has collected hundreds of live beach-stranded shells, placed them in safe habitats and watched them revive and move away. Also, depending on the circumstances of wind and tide, many stranded shells, as long as they do not dry out, are able to rescue themselves. I have watched such shells dig into the wet sand awaiting more salubrious circumstances.

Certainly over-collection of live shells is not the only reason for the decline of marine life. Pollution, dredging, seawalling, mangrove destruction, channel creation, inconsiderate boating (boaters who make "valleys" through the flats with high-powered propellors) contribute greatly to mollusk destruction.

We can't do much about some of those factors, but we can limit over-collecting. If California can do it, we ought to be able to do so too.

In the very cold winter of 1976-77, when the weather was terrible here for many weeks, and the rarest natural phenomenon we saw one winter's morning was a ten minute stretch of sunshine, many hundreds of large Horse Conchs, *Pleuroploca gigantea,* were exposed by extremely low tides. These tidal conditions were repeated day after day for the better part of a week. Hundreds of visitors crowded the beaches, invaded the Horse Conch's natural habitat and carted away many tons of these live animals. Gunny sacks, boxes, nets, plastic bags — all manner of gear was employed to get these animals out to the road and into the cars. Such pre-dawn shelling on a negative tide cannot be tolerated. It is a crime against nature and is unthinkable on a Sanctuary Island. Yet it happened. And for what? Within a day or two an overpowering stench undoubtedly set in, for most of these greedy people knew nothing of shell preparation. Rotting conch flesh and passengers in cars cannot coexist for long. Piles of aromatic conchs were dumped on the side of the road in north Florida and Georgia. I know this to be true, for a conscientious mother wrote me this from

See also Appendix II, the fine Code of Ethics of the American Malacological Union.

Cleveland, Ohio:

". . .Jimmy and Dale were almost persuaded to look and let live, but their father, the big lummox, joined them and my voice was lost in the turmoil. When we started to get into the car the morning we left Disney World, the stench was overpowering. We had to unload *everything* in the parking lot. The now very dead (and juicy) Horse Conchs were leaking from their plastic bags. Shoes, clothes, the trunk carpet, suitcases, paper boxes were all saturated with stinking rotting liquefied conch flesh. Also, most painful tragedy of all, Father's golf bag.

"The ride home was a nightmare. It was too cold to open the windows and too smelly not to. I am now trying all manner of things to rid the car of its aroma — all to no avail. I fear we shall have to drive it until its wheels fall off, for it certainly can not be traded."

Although it was just this sort of outrageous activity that brought the total ban on live shelling along the coast of California, it is perhaps unrealistic to seek a total ban on Sanibel. Of course, to admit this is to accept the fact that California and Ecuador are capable of more conscientious stewardship than we are. But in spite of the hundreds of letters written to me and the local paper expressing favor for a live shelling ban, when this controversy was at white-hot fervor, there are enough shortsighted money interests around to prevent it. Typically the large condominia and some motels advertise all over the country for visitors to come to our shelling beach in quest of beautiful "gifts from the sea." One such four-color ad photo, taken from the water, showed the condominium in the background. In the foreground was the beach on which was a pile of beautiful shells being examined by a bikini-clad mermaid. Close examination revealed that both Pacific and Atlantic shells were present, and clearly discernable inside one large shell, written with a felt marking pen, was "$3.75."

But it looks as though we are going to get some control. The Sanibel City Council is considering the following recommendation proposed by a group of concerned citizens who discussed the issues with Dr. F. G. Walton Smith of Planet Ocean and Dr. Gilbert L. Voss of University of Miami's Rosenstiel Marine Laboratory:

"1. That the City be urged to conduct a study as suggested by Dr. Voss as a means to provide factual base line data.

2. That the many instances of ignorant, greedy people loading baskets and buckets with hundreds of live shells brought up by winter storms makes us recommend as strongly as possible that the Sanibel City Council take a positive position at once on the subject of conservation as it relates to live shell collecting.

   Such a position should

   a. encourage respect for the mollusk as a living creature
   b. call attention to the value of mollusks as natural resources for their own sake as well as for their broad economic benefit to the economy of Sanibel.
   c. establish an emphasis on conservation education in relation to shell collecting equal to the city's emphasis on other conservation efforts.
   d. produce a definition of commercial shelling, with the purpose of controlling volume operations where desirable and indicated.
   e. pursue the erection of signs in appropriate public places to reflect the City's attitude.
   f. enlist the support of the Sanibel-Captiva Chamber of Commerce and the Hotel-Motel Association in an educational effort for their own benefit as well as for the island as a whole."

So all of our efforts to give some protection to Sanibel's shells *may*, just may, bear fruit. Why shouldn't we end greedy exploitation and at least establish reasonable limits to live shelling? A few decades ago, egrets were greedily slaughtered and their plumes sold into a frivolous though lucrative millinery trade. This very destructive practice was halted by foresighted conservationists. The same can be done for Mollusks.

21 *THE GRAND OLE OSPREY*

There are fifteen raptorial birds that appear on Sanibel, if you don't count the owls. The Osprey probably attracts the most attention of them all. Of course when our national symbol, our only pair of Southern Bald Eagles, produces young, there's considerable flutter among Sanibel naturalists. This happened for the first time in many years during the 1976-77 breeding season.

Many people are acutely aware of the existence of the Red-Shouldered Hawk, probably the most common hawk resident here. And in the fall and winter there are hundreds of Kestrels or Sparrow Hawks to be seen resting on electric wires all over the Island. About twice a year Pigeon Hawks or Merlins show up and the very rare and endangered Peregrine Falcon is sometimes sighted, as is the Marsh Hawk, the Broad-winged Hawk, the Red-tailed Hawk, the Short-tailed, the Cooper's, the Sharp-shinned and occasionally the Swallow-tail Kite.

Turkey Vultures and Black Vultures can often be seen behind the Island Cinema in the summertime, just before the show starts. They provide quite a show of their own. One May evening in 1977, while I was waiting for "Marathon Man" to get under way, I watched nineteen Black Vultures and eleven Turkey Vultures working over the garbage behind the nearby grocery store. It was quite an ornithological experience as well as an olfactory one.

The subject under discussion here, however, is one species, *Pandion haliaetus,* the Osprey or Fish Hawk, sometimes called the Sea Eagle — which is, by the way, the exact meaning of its specific name. Its generic name, *Pandion,* is the same as that of an ancient king of Athens.

This is an almost world-wide species which inhabits nearly all coasts and even large lakes and rivers of the great continents of the world. It has never been found in New Zealand, but that is about the only major area from which it is missing. It does occasionally show up in Hawaii, but it does not breed there. Otherwise it's found nearly everywhere — in Africa, Asia, North America, South America, Australia — and it even breeds on many oceanic islands. It used to be found in Europe in considerable abundance but it suffered greatly from human pressures and was wiped out almost completely in several countries.

Gamekeepers almost obliterated this species in Scotland and the few that remained were, in fact, completely wiped out by egg hunters. Collecting birds' eggs is a rather dastardly habit of many British bird fanciers. Fortunately this does not seem to be a great problem among Americans. Recently the Osprey has returned to Britain and is making a modest recovery.

The Osprey occurs as a single species in its own family, Pandionidae. There are some five races that vary slightly in coloration and size.

For many years there was a famous old Osprey nest in a dead mangrove on the "Ding" Darling tract near Station 9, at the bend in the dike road. This nest collapsed with age or weight in December 1976, but a few weeks later, after diligent labor, a large new nest was completed nearby. In fact, from the point of view of the visitor, the new nest is better, for it is in full view and easily observed. Presumably the same pair whose home had collapsed built and moved into the new nest.

In the 1974-75 breeding season, the female that had occupied the old nest made a crash landing and became ensnared in a mangrove tree and died. One of the youngsters managed to survive under the father's care. We believe the original male is still one of the pair that occupies the new nesting site at the present time.

Ospreys live a long time — perhaps forty

An Osprey bringing food to its young

years — and individual nests have been occupied annually for over four decades.

Dead trees are preferred nesting sites, for they offer better protection and visibility. But there are only a limited number of suitable dead trees on Sanibel for the Ospreys to nest in and as a consequence, they have sought out a number of artificial sites. For example, out in the main channel of Pine Island Sound can be found quite a number of Osprey nests — ten or twelve feet above the water, on channel marker posts. The birds become conditioned to rather heavy boat traffic with people stopping to take pictures while the mother is incubating the eggs or, later, when the chicks are being fed. They become quite accustomed to this sort of bother and successfully raise their young amidst a lot of human commotion.

On Sanibel the Fish and Wildlife Service provided three nesting platforms on the "Ding" Darling Refuge and all have been successfully occupied by Ospreys. Two of these are at sites where it's quite easy to observe the activities of the parents and young, both at nest-building and nest-repairing time and as they hunt for fish to feed themselves and their young.

Tragedy occurred at one site during the spring of 1977. Apparently the adult birds chose a dangerous piece of material to include in the nest — probably a piece of nylon fishing line. One adult became entangled and died. Its corpse could be seen hanging from the nest for several weeks. Whether the other parent raised any young at that nest I do not know, but young had hatched by the time of the tragedy.

During the nest-building time, one can see the parent birds stoop on dead Mangrove limbs, break them off while in flight and carry them away to the nest. Sometimes these sticks may be three feet long and must be quite heavy, but this powerful bird is able to carry substantial weight.

The female Osprey weighs about three and a half pounds and can carry more than a third of her own weight, so these birds are well equipped to handle heavy sticks, not to mention the rather large Mullet and Sea Trout or Weakfish that we see them feeding upon.

On sighting its prey, an Osprey may descend at great speed from considerable height, perhaps a hundred feet, with its talons stretched out ahead of the rest of the body. It hits the water with a tremendous splash, and sometimes completely submerges. A moment later it reappears with the fish clutched in its talons, after which it takes to the air, stalls for a second in mid-air as though its engine had stopped, and shakes its whole body briskly to rid itself of the excess weight of the water entrapped among the feathers. This enables the bird better to handle its burden.

The four-toed foot of this creature is constructed so that one front toe bends around to meet and work with the back one, enabling it to grasp a fish in its talons, with two toes on each side. The scales on the underside of the feet are very rough and might be said to be hooked or barbed. The sharp talons become deeply embedded in the fish, which is carried head foremost, for after all, fish are very streamlined creatures. To carry the fish tail first would leave fins flapping in the breeze and such wind resistance would impede progress.

It has been recorded that Ospreys have irrevocably locked onto fish that were too heavy to handle. When this occurs, the bird is lost, for the fish simply dives and the Osprey is drowned. It is an interesting fact that large fish have been caught with Osprey talons still in them from such an incident, the bird having decomposed, but the talons remaining.

We have had a great deal of conflict on this Island between the Ospreys and those who furnish our electric power. With natural sites being in short supply, the Ospreys have, nearly every year, taken to nesting on live utility poles that carry high voltage electricity. Ospreys almost always seem to choose those utility poles that have two cross arms; that is, those at a street corner or a place where the line is angled and added strength is needed.

In the nesting season of 1975-76, there were about five such nests on Sanibel, and the utility company was considerably upset with the "outages" caused during rainstorms when Osprey nests would create a short circuit that would blow transformers, fuses, etc. On one occasion an Osprey nest caused a fire that cost several thousand dollars. Not only is such a situation costly in terms of money and interrupted service, but it is dangerous to the birds themselves as well as to the electrical workers who are called upon to correct the problem, often in inclement weather. During the season of 1976-77, this writer worked closely with Lee County Electric and Homer Welch, who directs that Cooperative. We

sought a solution to the conflict that occurred between our much-loved bird species and those who supply our electricity.

A three-part program was developed:

1. A number of us banded together and raised money to erect a total of five artificial nesting sites on Sanibel using old discarded utility poles provided by the Electric Coop. A tried and proved design was furnished to us by Charles LeBuff and the Fish and Wildlife Service; Dr. William Frey, a Sanibel businessman, furnished labor and actually moved one nest that was too close to a developed area; a jury of ornithologically-oriented people was invited to participate in nest site location selection. Pat Hagan, of Fish and Wildlife; George Weymouth, a well-known local birder; and I were the only ones who responded and we agreed upon a number of sites. Then, with the help of funding from many citizens, we erected five posts and platforms on Sanibel. It is planned that for the future perhaps a dozen more will be erected.

Of the five that went up for the 1976-77 nesting season, two were immediately occupied with success, and interest was shown in two of the other three. So this facet of the program may very well develop successfully.

2. At the same time we were erecting alternate nesting poles, Lee County Electric Cooperative undertook procedures of its own. At the first sign of the first sticks to be deposited on a hot utility pole, they sent workers out to remove them. The idea here was to discourage the birds from building on these dangerous sites and remove the sticks before they were nests, and long before eggs were laid.

Photo, Peter Larson
Island Reporter

Osprey alighting on nest atop utility pole amongst hot wires carrying high voltages

3. The other function that the Coop undertook was to erect on the hot utility poles which the Ospreys seemed to prefer, some kind of device that would discourage their returning. This strategy has not worked out very well because small "windmills" that were placed on the poles froze up in the salt air when their bearings became corroded. Also, bent pieces of sheet plastic were placed in such a way as to, theoretically at least, discourage the birds from nesting. Rubber cones such as those used to mark lanes during highway repairs were nailed up on the utility poles. None of these methods worked very well. More innovative devices are being tested and developed.

In sum, we got a lot of cooperation from the Lee County Electric people and from the citizens of Sanibel and Captiva, but the Ospreys themselves proved to be most uncooperative and persisted in their efforts to rebuild on utility poles. As the very cold winter of 1977 progressed, Lee County Electric had more urgent problems to handle than those caused by a few Osprey nests, and during these weeks of considerable stress, three pairs of Ospreys did build on hot line poles. They got so far along it was deemed inadvisable to destroy those nests. We simply hoped that they wouldn't cause any trouble in rainstorms and we expected that by the next season we would have proper discouraging devices and additional artificial nesting sites available so that this unfortunate conflict could be forgotten.

For quite a number of years the most visible Osprey nest of all was the one atop the microwave tower of the Telephone Company.

Homer Welch devised this method of discouraging Osprey nesting. Shown are rubber road marker cones secured to a hot utility pole.

Photos, Laura Riley

This is what we provide in place of the hot utility pole. (Note head of incubating Osprey.)

This tower, which is the tallest structure on Sanibel, was topped with an Osprey nest which produced quite a few youngsters over the years. But in the nesting season of 1975-76 one youngster plunged to his death from the top of this very high tower. The Telephone Company has since erected a quite successful wire cone over the top of that pole and this location is no longer occupied by Ospreys. The Telephone Company also erected, some years ago, a fine alternate nesting site on their property but it has been consistently ignored by the Ospreys who sought the great height of the microwave tower.

The 1974-75 nesting season was rather successful on Sanibel with twenty-four nests which produced and fledged ten young. The next year there were twenty-eight nests and seven youngsters. For the 1976-77 season, there were thirty-seven nests and twelve young.

The Osprey picture on Sanibel has improved over that of years gone by, so we believe that the species is recovering after its years of problems created by biomagnification in its food chain of persistent chlorinated hydrocarbon pesticides. We expect to have the Grand Ole Osprey around for a long time.

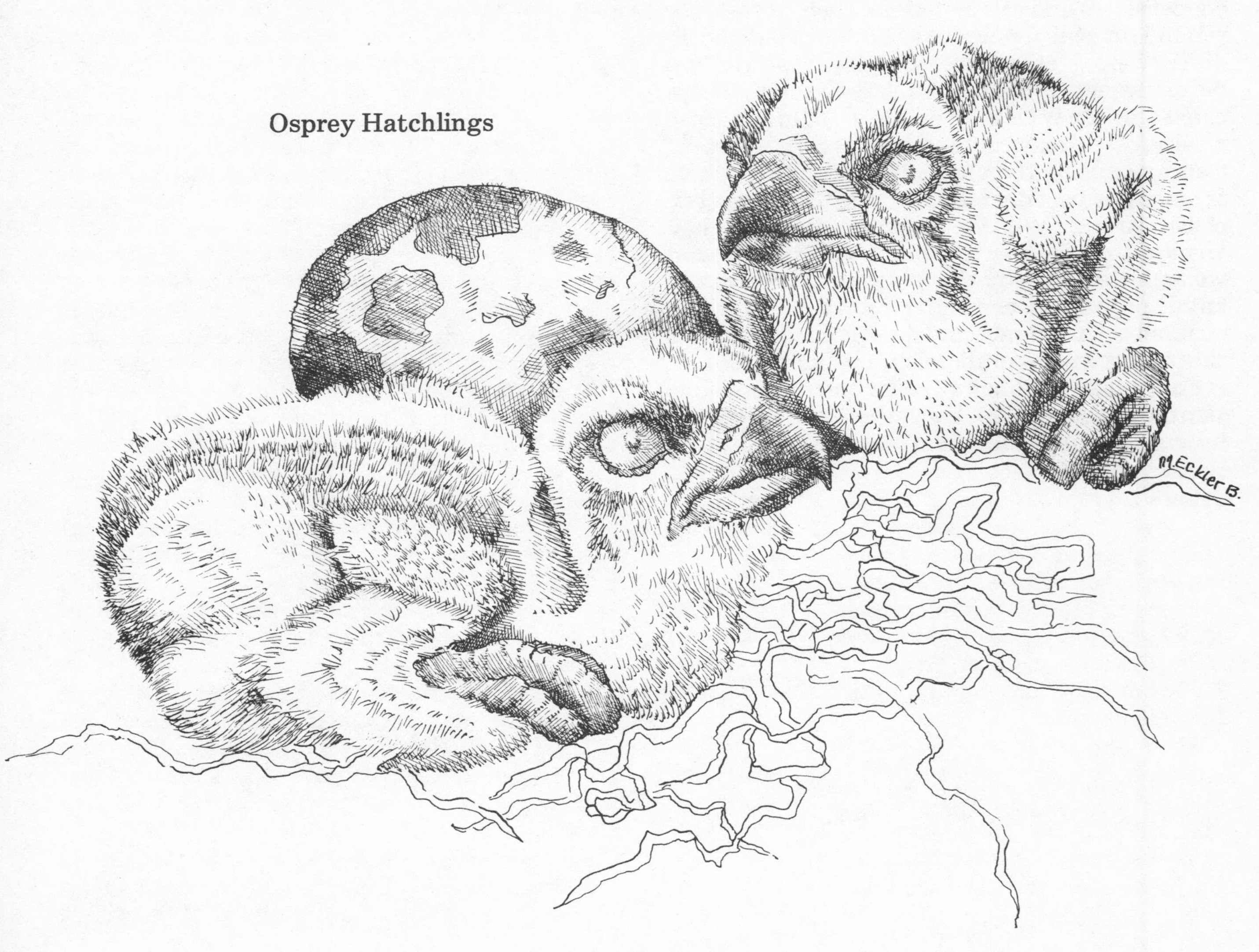

Osprey Hatchlings

22 *TO FEED OR NOT TO FEED* That Is The Question

Here on Sanibel many people feed the wild animals: the Raccoons, the Opossums, rodents, Alligators and many, many kinds of birds, including the two vulture species. I even know one young man who has "tamed" a Red-Shouldered Hawk to come to his home for handouts. Sometimes, as in the case of the vultures and rodents, this feeding is unintentional and is due to sloppy housekeeping around commercial zones. In many other cases it is intentional and often quite harmless. It is a fun thing to do which is, I suppose, the reason it is so widely practiced.

Tourists obviously enjoy the experience of feeding gulls on the beach at Tarpon Bay Road, or on the Causeway Islands, and at many other places where the gulls are conditioned to sky dive for bread and other food scraps. The gulls' ability to float almost motionless a foot or two above one's head offers the best possible close-up view of these beautiful creatures.

At least every other house on Sanibel has a bird feeder. Dick Beebe attracted and photographed many rare species in this way. Some residents spend considerable amounts of money on wild birdseed and sunflower seed. Others attract the rascally Raccoon and it is not uncommon to see "tame" Raccoons hanging around houses where they are accustomed to receiving free handouts. People often toss fish-heads and scraps to Alligators, a deplorable practice we have already discussed in a previous chapter. And of course every marina has its free-loading Pelicans awaiting daily fish scraps.

What about these practices? Are they good or are they bad? What about the effects on wildlife? What effect is there on those of us who live among the animals and have encroached upon their habitats?

I have discussed these matters with many Island authorities, including Charles LeBuff, Island zoologist; Glen Bond, "Ding" Darling Refuge Manager; ornithologist Dick Beebe, and Conservation Foundation Director Richard Workman. A kind of consensus has emerged which is here discussed.

The Raccoon, *Procyon lotor,* is fed intentionally by kindhearted people who deliberately attract the animals to their homes, and unintentionally by careless garbage disposal practices such as one sees around shopping plazas, grocery stores, etc.

There is really only one advantage to anyone's feeding of Raccoons, which we will touch on later. There are, however, quite a few disadvantages. If fed regularly, the Raccoons may not dump garbage cans, but when their population explodes and there is not enough food for them, you can expect raids on garbage and anything else that might prove edible.

Raccoons are being pressed into localized overpopulation situations because of land development. As their normal environment is being destroyed, they must go somewhere and human confrontation is inevitable. They often take the easiest and sometimes the only route and go to kind-hearted feeders who artificially maintain these often sick animals.

The Raccoon not only carries feline distemper, but canine also. It is well known that all Procyonids, including the Panda, the Coati and the Kinkajou, when held in captivity, must be protected against common canine *and* feline diseases.

Feeding coons makes these animals dependent upon man for their food and they lose their ability to earn their own living. They also lose their fear of man and thus become dangerous to children who too often hand feed them.

Sick Raccoons can often be distinguished by a characteristic staggering gait. Although they probably do not often transmit disease to humans, there are records of Raccoons carrying rabies. Raccoons can carry this disease without showing any symptoms. So really I suggest it is much better to leave this animal alone. Raccoons, concentrated unnaturally because of habitat destruction or simply by human invitation, are more likely to be diseased because they live in closer contact with one another than is normal and often receive an improper diet.

I know of one man's "pet" coon population that is fed old white bread. This can still be

A member of Sanibel's nocturnal sanitation crew

purchased for nickels and dimes at wholesalers' in town. Large supplies are brought here. As everyone knows, this fluffy white stuff is hardly a proper diet. The wonder of it all is that the beasts are able to survive. If you *have* to feed coons, feed a good quality dog or cat food and at least your friends should be healthy.

There is one advantage to feeding the coons, and this demonstrates the unexpected complexities that emerge when we try to "manage" a wildlife sanctuary such as Sanibel: Charles LeBuff tells me that here Raccoons no longer destroy so many Loggerhead Turtle nests as they did. He believes that all the artificial feeding has encouraged the coon to change his ancestral feeding habits to exploit the garbage can or the soft hearts of the ever-increasing human population, and is thus putting less pressure on the Loggerheads. Lending substance to this theory is the fact that on turtle nesting beaches that are uninhabited by people the coon is still a major Loggerhead nest predator.

Other "fed" mammals present problems. The Opossum causes less concern than the Coon, but many of the same principles apply to this species as well. Two species of skunks have been recorded here. Although rare today, they could become common; and statistics show that elsewhere the rabies threat from them is ever so much greater than from the coon. Careless feeding resulting from poor housekeeping can give a competitive advantage to the two exotic Asian rats over our own desirable indigenous rodents. Another problem relates to the feral domestic cats which should either be *really* adopted (and this includes a proper home, diet and veterinary care) or they should be humanely destroyed.

Dick Workman has this to say: "Recent studies by competent wildlife biologists have shown that any feeding of wildlife, including birds, can have an unnatural and detrimental effect on their populations. It enslaves them to an area. When stopped, it is traumatic to the wildlife and can cause an immediate collapse of an artifically built up population."

Workman's comment is rather brave, I think, because he includes birds, and nearly everybody on this Island loves to feed the birds.

Certainly if you are going to feed birds it should be done regularly and not suddenly stopped. If one is not prepared to do a good job of it, it shouldn't be started in the first place. There appear to be periods, especially during breeding season, when birds can become dependent upon artificial sources of food.

Bond states that insofar as bird feeding on Sanibel is concerned, it isn't so much the food that's important once you start, as the fresh water. Stress periods for birds here come mostly during times of drought. When natural water disappears, birds appreciate and need the provision of fresh water drinking facilities. Elsewhere, especially in the north, stress periods for birds usually result from ice storms or heavy snows which cover natural food supplies. In such situations, the feeding of birds can help starving populations.

Dick Beebe states that many species of birds can now stand the rigors of northern winters because of artificial feeding. The migration habits of some birds are changing and some now winter in areas of rigorous climate where only a few years ago they migrated to warmer latitudes. An excellent example of this is the Canada Goose which not long ago wintered in the southern states in the Mississippi Valley. Today, due to an artificial food supply in refuges in Wisconsin and Illinois, the migration pattern has changed and this species no longer comes south.

So in deciding your own course of action in regard to the complex subject of feeding the wild animals of Sanibel, let your conscience be your guide. But one would hope that your conclusion would not involve the perpetuation of a population of sick feral cats, the unnatural concentration of not-so-healthy Raccoons, and the creation of dangerous Alligator situations through thoughtless (not to mention illegal) feeding and "taming" of these interesting animals.

For those persons who spend only the winter months on Sanibel, I suggest that before making the decision to adopt whatever engaging creature captures your fancy, you give some thought to the fact that your December "child" will be April's orphan.

23 *THE KIRTLAND'S WARBLER ISN'T HERE BUT—*

During Spring and Fall migrations, Sanibel boasts many Wood Warbler species. The Ovenbird, Palm Warbler, Northern Prairie Warbler, Pine, Blackpoll, Bay-breasted, Yellow-throated, Blackburnian, the Black-throated Green, the Myrtle, the Black-throated Blue, Cape May, Magnolia, Yellow, Parula, Nashville, Orange-crowned, Tennessee, Blue-winged, Golden-winged, Worm-eating, Prothonotary and the Black-and-White Warbler are all known here, many in very large numbers. We also have the Florida Prairie Warbler as a permanent resident. During the migratory seasons, many Sanibel birders can be observed whispering to the Warblers, thus calling them into view.

The Jack Pine or Kirtland's Warbler, *Dendroica kirtlandii,* has never been recorded on Sanibel. If found here it would be a stray, but my ornithological friends tell me that this is possible. So, based on this rather lame excuse and because I have had some small experience with this species in my lifetime, I'm going to stretch a point and discuss this fascinating bird and its unusual history.

Molly Eckler Brown has done an illustration of the Kirtland's Warbler. This time her work was not done from life because that would be practically impossible. There are so few Kirtland's Warblers left that they might fit in your hat.

Although every bird watcher in the United States who can see anything smaller than a Pelican has heard of the Kirtland's Warbler, there are some facets of its nature and history that are not too well known.

It was first discovered on May 13, 1851 on his farm near Cleveland, Ohio, by Dr. Jared Potter Kirtland, a physician and one of Ohio's great naturalists. The describer of the Jack Pine Warbler, and the man to whom Kirtland sent his specimen, was Spencer Fullerton Baird of the Smithsonian Institution, then considered to be the best-informed man on the subject of North American vertebrate fauna and the author of many works on birds and other animals.

Today there are fewer than 500 of these little birds, but their association with the Jack Pine, *Pinus banksiana,* and with their parasite, the infamous Cowbird, *Molothrus ater,* and also with the late notorious Nathan Leopold of the Loeb and Leopold murder trial, adds to the fascination of this species.

My own interest in this animal was first stimulated back in 1949 when as a young man I lived on the island of Eleuthera in the Bahamas, an island less than three hundred fifty miles east of Sanibel.

As a somewhat nearsighted ground-staring herpetologist, I didn't know much about the avifauna of Eleuthera until one day a giant of a man — both physically and in his profession — showed up and wanted to be directed to the Kirtland's Warbler's wintering area. Seven-foot-tall Josselyn Van Tyne, head of the Bird Division at the University of Michigan Museum of Zoology in Ann Arbor, was one of the best known ornithologists of his day.

He took his .22 caliber "shotgun," really a .22 rifle loaded with fine pellets rather than a single lead bullet, and together we hunted the scrub lands of Eleuthera in an effort to shoot specimens of this already rare bird. The experience with that great man was my first exposure to the avid "study skin" collecting activity that was the method employed by many ornithologists to amass museum collections for future study. Today, more ornithologists shoot with camera than with gun, and this is far better, to my way of thinking. I must say that I was pleased that Dr. Van Tyne's shooting expedition on Eleuthera, in which I participated, was a failure.

Much later I visited the University of Michigan's birdskin collection, which numbered about two hundred thousand specimens, including such unbelievable rarities as the Passenger Pigeon, the Great Auk, the Eskimo Curlew, the Carolina Parakeet and the Whooping Crane. Such huge collections may seem a bit too much, but when one considers that they were gathered over a long period of time, many dating back to the last century, and that they have been the basis of studies by conscientious ornithologists who have produced the data we all use in today's conservation activities, such a collection ceases to be so distasteful. Collecting by professional ornithologists has probably not contributed to the demise of any bird species. Those that have become extinct by man's agency have been pushed over the brink by more lethal factors such as habitat destruction (as in the case of the Ivory-billed Woodpecker), or by overshooting (good examples of which are the Passenger Pigeon and the Carolina Parakeet). Over-exploitation for food did in the Great Auk and the Dodo.

Today, professional ornithologists do much less collecting of specimens, conduct many more live studies, use photography as a tool in place of the gun, and undertake behavioral field studies of a nature that is harmless to dwindling populations and in fact supports them, as we will see later in this discussion.

The Kirtland's Warbler winters on the island of Eleuthera as well as on several other major islands in the Bahamas. They can be seen there even now, but one must search diligently. Richard Beebe and I sighted a specimen at Palmetto Point, Eleuthera, in October, 1975. It was sharp-eyed Beebe's identification, and he doesn't make mistakes. This was the first confirmed Bahama sighting in many years and was widely publicized.

Actually the problem with the Kirtland's Warbler is not in the Bahamas nor is it that a stray might find its way to Florida. The real threat is a relatively new invasion of its habitat by the parasitic Cowbird which threatens its breeding success. Kirtland's has evolved no defense against this avian parasite which lays its eggs in the Warbler's nest and depends upon the Kirtland's to incubate its eggs and raise its young.

Another basic threat is habitat reduction. Jack Pine forests where Kirtland's breeds in the summertime are in the northern part of the Southern Peninsula of Michigan, where various Audubon Societies are monitoring the annual production of nestlings and where they finance the trapping of Cowbirds to lower the parasitism. Controlled Jack Pine forest fires are purposely set in order to create the necessary habitat for the Kirtland's, for available habitat is very limited in area.

Many forests in that part of Michigan have overmatured Jack Pine habitat. Kirtland's only accepts small young Jack Pine forest breeding areas. Hence those controlled fires. New habitat is created by forest fires which stimulate Jack Pine seed dispersal and germination and insure the proper succession.

In June and July 1976 there was considerable elation among the Kirtland guardians, for the breeding population was up 20 pairs from the previous year, to 199. By August, 1976, this elation had turned to ashes. The Michigan National Guard had laid down artillery fire in the area and accidentally burned down 335 acres of prime young Jack Pine habitat. The trees had been just of the right age to attract the breeding pairs, of which there were six. Whether any of those adults were destroyed is not known for sure, but in any case the breeding success in those six nests was frustrated.

According to the Michigan Department of Natural Resources, the "habitat was set back 15 years." Or more, it seems to me, if you consider that the young pines were immature and as a result of the fire, could not reseed as do adult trees.

After Dr. Kirtland's 1851 discovery of the species, it was realized that a specimen had been mouldering at Harvard's Museum of Comparative Zoology since the early 1840's. This one had been taken on a ship near the Bahama Islands. The literature shows that over the years (mostly before 1920, during the heyday of this species) 71 have been collected from the Bahamas winter quarters. I don't believe that any actual shooting of this species is taking place anywhere today, so any specimens taken are accidentals that have died from other causes, such as the example recently found near Cincinnati, Ohio.

Roger Tory Petersen calls the Wood Warblers the "butterflies of the bird world". Kirtland's Warbler is one of the most attractive. The male is bluish gray above, with a striking yellow breast. The back is gray, heavily marked with black bars. It has a black face, interrupted white wing bars and dark streaks on its sides. The female lacks the mask but is similar in other respects, though not quite so brightly colored.

Some observers feel that the Kirtland's Warbler has the most beautiful song of the warblers and it is thought to be somewhat like a mixture of the sounds of the House Wren and the Northern Water Thrush.

For a bird bearing the name of such an illustrious citizen as Dr. Kirtland, and one that has been studied by the great ornithologist Van Tyne, it seems to me ironic that the notorious Nathan Leopold would also have had a hand in the revelation of this animal's life history. I refer to the same Nathan Leopold who was co-defendant in the "crime of the century," the Loeb and Leopold case of the twenties in Chicago.

Before Leopold's incarceration in Joliet Prison for murder, he made two visits to the Michigan nesting grounds of the Kirtland's Warbler in the summers of 1922 and 1923. He organized both expeditions, having become interested in the bird when he was a student at the University of Michigan. On the 1922 trip, he was unsuccessful in finding the bird, but in 1923 he located it, took notes, made pictures and accumulated enough material to justify an article published in *Auk,* one of the most prestigious ornithological journals in the world.

Clarence Darrow, the famed lawyer who was the defense attorney in the Loeb and Leopold murder trial, said that Leopold would be imprisoned for life if he didn't receive the death sentence. Long years after Darrow's death, during World War II, Leopold was still at Joliet Prison in Illinois where he undertook a cooperative research study in malariology. His studies led to the startling realization that side effects experienced after the ingestion of antimalarial drugs of the group known as primaquines were due to hereditary factors in the patient.

These and other important scientific contributions by this murderous genius won him his parole after World War II. Leopold spent the rest of his life in Puerto Rico, working in the health care field. He never forgot his interest in ornithology and not long before his death, he published his "Checklist of Birds of Puerto Rico and the Virgin Islands." This document stands today as a leading ornithological contribution for that region.

So the saga of the Kirtland's Warbler is woven among many kinds of personalities. I hope the final chapter of this species' history is not soon going to be written. I hope that the Jack Pine forest regeneration necessary for the perpetuation of this animal will take place. I believe that the Audubon Societies of Michigan will continue to fund the Cowbird studies and trapping programs and that the Bahamas National Trust will continue their support and work so that the Kirtland's Warbler can be around for a long time to come.

Maybe one will blow in here sometime.

* * *

Postscript

As I sit on my patio trying to think of a reasonable excuse for including the fascinating story of the Kirtland's Warbler in a book on Sanibel's flora and fauna, I have just witnessed one of the world's greatest ornithological sights.

Today is April 3, 1977. The time is noon. Suddenly, an hour ago, the sky behind my home, and over the freshwater pond that is there, became darkened by tens of thousands of swirling Tree Swallows, *Iridoprocne bicolor.* They swirled above the pond and above my house in clouds reminiscent of the great locust swarms of Asia and Africa. One cloud flew down into my back garden. They descended in such a way as to appear almost to be a cyclonic cloud reaching to the ground. Tens of thousands of these birds landed on three Southern Bayberry trees, *Myrica cerifera.* There were so many of these tiny vibrant birds each of which could hardly weigh an ounce, that the trees were bent double by their combined weight, which must have been hundreds of pounds. A great fluttering took place among the trees for forty-five seconds or so, and then the birds rose, after having stripped the trees of the wax-loaded berries, for our Myrtle belongs to the same genus as the Wax Myrtle from which our ancestors made those fine aromatic wax candles.

The tens of thousands of these truly beautiful birds with their lovely iridescent green backs again rose and swirled above the pond, only to descend once more on three more trees. All of this took place only twenty-five or thirty feet away from me. The rustle of wings, the liquid song, the obvious tremendous weight of accumulated thousands all add up to an unforgettable memory that certainly rivals in interest, to me at any rate, the sight of more than two million Flamingoes that one can sometimes see at one time on Lake Nakuru in Kenya.

The Bayberry wax obviously has food value, for these otherwise insectivorous birds are surviving on this diet. In 1975 and 1976 we saw them in December, January and February, doing the same thing. This is the first time I have witnessed them in the Spring and so close up. Obviously they are congregating for the return north.

This species normally nests in the north in hollowed-out cavities in trees but takes very readily to nest boxes. I'm told by friends from New England that, just prior to migrating south in the fall, the same sort of congregations take place as they feed on the Northern Bayberry, *Myrica pensylvanica.*

I feel very fortunate to have had this really close-up experience of this lovely bird.

24 *SANIBEL'S NOCTURNAL GLEE CLUB:* Our Toads and Frogs

There are twelve kinds of frogs and toads found on Sanibel. For a barrier island lying several miles out in the Gulf of Mexico and subject to the buffeting of hurricanes, this represents a rather rich herpetofauna. The toads and frogs, except for one species, depend largely on the fresh-water central wetlands which are today unique to Sanibel for no other South Florida barrier island has such a system. Palm Beach and Miami Beach islands used to have central fresh-water systems, but these have long since been destroyed by human development.

Of the twelve species which we shall discuss here, three are introductions and nine are indigenous. Of the exotics, one — the Marine Toad — has been seen only once. But due to its prevalence elsewhere in Florida, it can be expected to occur here in great abundance before too long.

Let us consider first the nine native species and see what is happening to them. They probably came here, and their populations were probably replenished with a certain frequency, before the days of herbiciding the water hyacinths on the mainland rivers. Some years ago, great rafts of water hyacinth (*Eichhornia*) used to drift down the Caloosahatchee River, and often acres of them would be washed ashore on the various islands. With these plants came many of the frogs and reptiles from the mainland. They were able to survive because of our wetlands system.

Today the Lee County Mosquito Control District is using herbicides on water hyacinths and also has tried biological control of the plant, using a beetle from South America. As a consequence, it's almost unheard of for a water hyacinth to be washed up on the Bay beach or around the Lighthouse of Sanibel today. If amphibian species suffer a bad year due to salt water intrusion on Sanibel, it is unlikely they will be replenished by the old method of transport via hyacinth raft. Of course hyacinths are exotic themselves, not having been in primordial Florida, so it's entirely possible — in fact, probable — that many of the herptiles that we now list as "indigenous" to Sanibel really did not occur here in the early days before the hyacinth. So with these factors in mind, let's discuss what we now have and what can be done to preserve them, for many are indeed contributors to the pleasure of living on, or visiting, this Island and their abundance significantly contributes to the control of undesirable insects.

There are three little native tree frogs — the Green Tree Frog (*Hyla cinerea*), the Squirrel Tree Frog (*Hyla squirella*) and the Little Grass Frog (*Limnaoedus ocularis*). The latter tiny frog used to be considered a member of the genus *Hyla,* but has recently been renamed as above. This species may be considered to be the smallest frog in the United States, usually no longer than 5/8 inch, and the largest "giant" on record only 11/16 of an inch long. While the Little Grass Frog is plentiful on the nearby mainland, it's almost impossible to find on Sanibel today. I believe this to be due to salt water intrusion, which is becoming an ever greater problem for several reasons: (1) the many perforations of the seal or aquaclude of the internal fresh water slough system, and (2) especially the leakage of high spring tides through the water control structure at Tarpon Bay, which allows entirely too much salt water into the central part of the Island. This salt water intrusion destroys amphibian breeding habitat.

The Green Tree frog is somewhat larger. It grows to 2¼ inches in length. It is usually a bright green, but can change color and may be yellow and sometimes brown. It has a beautiful white or cream colored racing stripe down its side and can easily be identified by this characteristic. In the wilder parts of Sanibel, the Green Tree Frogs can be heard in chorus from late May, throughout the summer, and I even have several records of the early days of October. When one is in the midst of several thousand "rain frogs" chorusing together, conversation is all but impossible. The bell-like call, while very melodic, is absolutely overpowering.

The Squirrel Tree Frog is somewhat smaller than the foregoing species, seldom exceeding an inch and a half in length. It changes color and has many variations and patterns — from brown to green and sometimes spotted. And usually there's a darker spot between the eyes. There may be a light line along the body, but this is much less pronounced than that of the Green Tree Frog. The name comes from its rather raucous call — somewhat like that of a squirrel, but perhaps more like that of a duck. Its call is harsh, repeated very frequently in chorus, perhaps once a second.

Both of these beautiful native species are considered highly desirable, due to their insectivorous habits and their abundance on the Island. Sometimes during a rainy night, literally thousands can be seen around lights along shopping center windows or on the front of your house if you leave the outside light on. They're very abundant but they are becoming less so. The same factor we mentioned for the Little Grass Frog — salt water intrusion — is affecting its breeding habits. It is hoped that we can correct this in time to save these and other species. But perhaps the most serious threat to both of these animals is the hated (and rightfully so) Cuban Tree Frog (*Hyla septentrionalis*), the giant of this genus. The females grow to be over five inches in length and the males attain as much as three and one-half inches. This animal will swallow anything that moves which is small enough to be crammed down its throat using both fists. This includes the native Tree Frogs, and it is my belief that the native species are suffering because of predation by this foreign interloper.

The Cuban Tree Frog can be readily distinguished from our natives because of its large toe discs and its skin, which is warty — not smooth as in *squirella* and *cinerea.* It is often brown; sometimes there's a pattern on the back that can be a light grey or pale yellow.

The Cuban Tree Frog came here with potted plants brought in for landscaping. Wherever there is a new, fresh landscaping job with potted plants from off the Island, this frog is soon evident. It is a rather domesticated creature, living in cisterns, rain barrels and fish pools and taking moisture from the axils of Palm Trees, the so-called Traveler's Tree, Bananas and especially bromeliads.

We know that the smaller Green Tree Frogs and the Squirrel Tree Frogs are attracted to the lights of buildings on damp nights and so are the Cuban Tree Frogs. We are thus likely to find as time goes on that our own native forms become scarcer.

I recently put one Cuban Tree Frog, eight Squirrel Frogs, one Green Tree Frog, and one Greenhouse Frog (a little animal which will be the subject of later discussion) in an aquarium, locked the door and left them. Between June 24 and July 8, the Cuban Tree Frog ate seven of the Squirrel Frogs, the Green Tree Frog and the Greenhouse Frog. There was still one little Squirrel Frog in with him which had learned how to hide on the bottom of the aquarium below the moss.

I realize that this kind of experiment doesn't prove that the larger species *does* actually consume the smaller ones in the wild. But it certainly is an indication in that direction, and such disruptions are not at all uncommon after the introduction of an "alien" or exotic animal into an already existing environment.

Another nasty thing about the Cuban Tree Frog — it is quite poisonous. The mucous from its skin can cause temporary blindness, and if the mucous gets into a scratch or hangnail, it can cause acute pain. One really should not pick up this animal because of its ability to cause very serious discomfort.

One of the most interesting of our frogs is another introduction, the Greenhouse Frog (*Eleutherodactylus planirostris*). This is a tiny little creature, seldom over an inch and a quarter in length. It came here, probably, from Cuba but it is also known to occur in the Bahamas. When I lived on the island of Eleuthera, for a while I thought that *Eleutherodactylus* was named after that island because it occurs there abundantly. But come to find out, *Eleutherodactylus* refers to the unwebbed free toes and fingers of this animal. The island was so-named because it was the New World's first free democracy.

These little frogs belong to a family known as Leptodactylidae that contains many species all over Latin America. The present species came to this Island by means of potted plants grown in the Miami area where this animal has been naturalized for at least forty years to my certain knowledge. An interesting thing about this species is that, unlike virtually all other amphibians, it does not require fresh water in which to breed. It lays its eggs in damp leaf mold and the larva undergoes complete metamorphosis within the egg. Only after the tadpole has got his hind legs and his front legs and has absorbed his tail does he break loose from the egg. It's a most unusual thing to watch and it's an easy process here on Sanibel to find these little animals because wherever there is a damp shaded area, with flower pots and leaf mold about, when you shuffle around or pick up a pot, the fast jumping cricket-size creature that is seen is in all probability a Greenhouse Frog.

Although most exotic introductions are harmful, I can't see prejudicial potential in what the Little Greenhouse Frog is doing to our environment. It simply contributes to the consumption of insects — probably a desirable function.

Another interesting frog that occurs on Sanibel but is seldom seen, is our largest native frog, the Pig Frog, *Rana grylio*. It is hard to see him, but he can be heard nearly every night in the summer. You can't mistake the call of the Pig Frog because it sounds just like what its name suggests — a pig grunting. Sometimes it is called Florida Bullfrog, for it is almost as large as the bullfrog known from other regions which furnishes that gourmet's delight, frog legs. In fact the legs of *grylio* are eaten in Florida. (However, most commercial frog legs consumed today in the United States are raised near Bombay, India, and shipped here frozen.) Here on Sanibel we do not have a great population of them, thus the deafening roar that is present where hundreds are breeding in a pond cannot be heard on Sanibel. Most years you can hear singletons or twos and threes at almost all parts of the freshwater wetlands system of the Island, during their summer breeding season. In 1975 it was a different story — with the Pig Frog and most of our other frogs and toads too. The weather was so dry that I'm afraid many of our toad and frog populations were decimated.

In the central part of Florida, when a drought occurs, the only problem that faces the amphibians is dessication. On Sanibel, there is a two-pronged threat: dessication and salinization. And most frogs are noted to be intolerant of much salt water.

All during the months of May and June and the first part of July, 1975, I vainly drove the roads and tramped the wetlands of Sanibel seeking a Pig Frog, to act as a model for a Molly Eckler Brown drawing. Areas that

Photo, Dallas Kinney, *Island Reporter*

Cuban Tree Frog. This alien invader is consuming our native frog species.

should normally be wet and flooded in June were dry, and well into July they were still dry. Many areas that still had dampness were very saline. The Tarpon Bay mosquito control weir is defective and the high spring tides carried large quantities of salt water into our fresh-water slough system by way of the Sanibel River. Frogs of all kinds suffered serious decline.

Then came some rains in mid-July, and on the night of July 18, during a one-inch rain storm, I studied Rabbit Road for signs that would tell me what was happening to our amphibian populations. Instead of hundreds of Green Tree Frogs, Squirrel Tree Frogs, Narrow Mouthed Toads, Southern Toads and Leopard Frogs, which one should see on that road on a very rainy night in July — I saw only dozens.

It is a sad situation, and a deplorable by-product of the drought and inadequate water control, because the frogs and toads of Sanibel are among our most desirable animals. They provide control of mosquitoes and other insects, in addition to giving much pleasure to the musically inclined members of our community.

(I hope that some of the efforts now under way by the Sanibel-Captiva Conservation Foundation and the City of Sanibel will help to control, maintain and preserve the fresh-water system of this Island, because this system is absolutely essential to many, many of our wildlife species.)

My wildlife expedition on Rabbit Road on the wet July night did, however, net the sought-after prize, a fine example of an adult Pig Frog. The record size for this species is six and three-eighths inches (from the point of his tail to the point of his nose). Of course if you stretched out his legs, he would be much longer than that.

Pig Frogs are brown or olive green and they have numerous dark spots scattered over them. The underside is white, or sometimes yellow, and the legs have some black markings. Farther down on the legs there are light spots.

This is definitely an aquatic, fresh-water species that lives in the marshes on the mainland around cypress areas and water lettuce lakes and areas where water lilies grow. On Sanibel it lives in the central slough.

It belongs to the typical frog family, Ranidae. These are usually long legged, smooth skinned and good strong jumpers and swimmers. The digits of the hind feet are webbed, while those of the front feet are not.

The genus *Rana* occurs throughout the world. The largest one of all is *R. adspersus*, the ridge-backed frog of southern Africa. This animal is so big that it reaches the size of a dressed-out turkey. It has two great snaggle teeth in its lower jaw and can swallow anything from a small rabbit to a pullet. It is a formidable animal indeed.

Cuban Tree Frogs in amplexus, the mating embrace of toads and frogs

I vaguely remember a fairy tale about a prince who was turned into a frog or a toad and only a kiss from a beautiful princess would cause him to revert to his princely status. I hope the wicked witch selected a more attractive species than *R. adspersus* to present to the fair princess.

One species, *R. esculenta,* is the main prey of the European stork, the famous bird that used to occur in large numbers in Holland, Belgium, Germany, France and nearby countries and figured in so many Western European folk tales. *Rana esculenta* was very nearly wiped out by the use of DDT in Europe and as a consequence the stork was nearly destroyed too. Today storks and *Rana esculenta* can be used as a sort of political or perhaps economic barometer: storks and frogs are found in the Eastern European countries where little money was spent for insect control during the heyday of DDT. Also, a few can be seen in backward Portugal where DDT was never widely used. But in the prosperous middle part of Europe, that is in the Common Market countries, very few storks remain, as do very few *Rana esculenta.*

The Pig Frog hasn't always been here on Sanibel. In 1952, Refuge Manager "Tommy" Wood brought in 39 specimens divided almost equally between the sexes and released them into what was later to become the Bailey Tract of the Wildlife Refuge. They seem to have prospered and have proved to be an interesting addition to our fauna.

I have been having a little trouble classifying such an animal as "exotic," for these animals are native on the nearby mainland. But even when such an animal is transferred artificially a short distance, it attains some of the character of a true "exotic."

The Pig Frog breeds from March through September and the larvae or pollywogs undergo a long period of development sometimes leading into the second or third year. They grow to be quite large, perhaps four inches in this species. Charles LeBuff found this species to be somewhat salt tolerant when, in 1964, he found some in water that tested out at 40% of sea strength.

The Pig Frog consumes crustacea, fish, other frogs, small snakes, small mammals and insects. In fact, the Pig Frog will swallow a pebble if you roll it in front of him. It will eat anything that moves that it can cram down its throat, using its two arms in the style made famous by King Henry the Eighth.

It also has its enemies: water snakes and garter snakes will feed on the larvae and the young, and alligators will consume their fair share, as will turtles, herons and raccoons.

The next animal we will consider is the Giant Toad, *Bufo marinus.*

It has been proved over and over again that when man introduces an alien animal into a new territory, environmental damage almost invariably results. There are many examples of this: the Mongoose of India, which has killed off most of the birds of Hawaii and much of the fauna of Jamaica, Trinidad, Puerto Rico and many other places where it was carelessly introduced to control snakes; the Giant African Snail, which destroyed the gardens of south

Florida a few years ago; the rabbits and the cactus in Australia; the more than seventeen major mammals which have been introduced to New Zealand, all of which have wrought untold destruction to the natural habitat of that beautiful land.

Florida has more than three hundred introduced animals from all parts of the world, everything from parrots and monkeys to Walking Catfish. One of Florida's most destructive introductions is known on Sanibel from just one specimen. But just wait — I believe the Giant Toad, *Bufo marinus,* will be here in abundance all too soon.

Bufo marinus is one of the biggest toads in the world. It looks a lot like the little one we see around here which is native to our Island, *Bufo terrestris,* but *marinus* grows to be eight inches long and can be more than three pounds in weight. One of these beasts can produce more than thirty-five thousand eggs per year and the tadpoles are voracious feeders. The young toads are larger than the young of our native species and consequently fare better and soon take over the whole toad environment.

They do untold damage to native amphibian populations. They'll eat anything that moves, which means they have some good habits, since they eat grasshoppers, palmetto bugs and other insect pests, as well as grubworms, mice and rats. Unfortunately, they also eat small birds, fish, small turtles and small snakes.

In Dade County Giant Toads have learned to eat dog feces. In Coral Gables there is a large thriving population that subsists entirely on this unusual diet. In the piney woods section of South Dade, *Bufo quercicus,* the Oak Toad, has all but disappeared, as has *Bufo terrestris,* both eaten up by this alien intruder.

Bufo marinus has been introduced to every country where sugar cane is raised. In Jamaica, it did away with many of the cane beetles that plagued the sugar population, but it destroyed many of the other animal species that normally live in Jamaica. The same is true of Puerto Rico, Haiti, Guam, Hawaii, the Solomon Islands, the Philippines, Central America, Trinidad and especially Australia, where its spread is so feared that in some areas a large bounty is paid for a single dead toad.

When I was a student at the University of Puerto Rico, I recall seeing literally thousands upon thousands of these animals on the campus at Rio Piedras. Car wrecks have been caused by *Bufo marinus* in Haiti on wet nights when they are so numerous on the roads that tires can get no traction.

In Dade County's Key Biscayne area they are very abundant. The zoo there has thousands of them living near the ponds and around the refuse dumps where they consume rats, mice, insects and almost anything else they can shove down their oversized gullets.

I once had a pair of Key Biscayne *Bufo marinus* in my collection. They were named Trisha and Edward and they were absolutely huge, each weighing more than two pounds.

The Giant Toad is dangerous and offensive in another way. If you squeeze the large parotoid gland on the head, you can get a jet of poisonous fluid that can squirt a distance of five or six feet. A dog may die if he mouths this animal. If you get this fluid in your eyes, it will sting and burn.

The Marine Toad, as it is also called, has been spread around Florida by the nursery trade. Nurseries usually have an abundance of water — often ponds, where toads can breed, and they also have cool damp hiding places — tin cans and so on, with moss and other features favorable to toads. This animal will often hide or bury itself in a pot and be moved long distances and subsequently appear in new regions.

Other amphibians that can be found on Sanibel are the Southern Toad, *Bufo terrestris,* which is the common toad of the South and the true toad still most abundantly seen here. It has a shrill musical trill, rather high, and breeds from April through June on Sanibel. Thousands of juveniles often appear toward the latter part of July or early August when the fresh water conditions are right, and by mid-September, they will have grown to about an inch. It is a valuable insect-eating animal, still common, and will endure as long as we have abundant good fresh water or until the Marine Toad wipes it out.

The Southern Leopard Frog, *Rana utricularia,* occurs on Sanibel and for some reason or other grows to be much larger here than elsewhere. We have seen them up to four and a half to five inches while usually, in other parts of their range, they are only about three and a half to four inches long. This animal breeds in the fresh water sloughs from April and May right on through the summer. I have heard a chorus as late as mid-October. The call is a scraping sound that's rather hard to describe, but quite distinctive.

The Narrow Mouth Toad, *Gastrophryne carolinensis,* is numerous and can be heard from April and May on into early October. It has a very distinctive voice that sounds like the "baa" of a sheep. It grows to be about an inch and a quarter in length, but can easily be distinguished because it's a short fat stubby little toad with a very pointed head and a very narrow mouth. A good place to see and hear this creature is along Bailey Road on a warm summer's night, during a slight drizzle. You will think you're either on a sheep farm or in an electronics laboratory where doorbell buzzers are being tested. This species should be especially interesting to Sanibelites and we should all encourage it because it feeds very largely on ants. We have so many very unpleasant ant species on this island that certainly we should encourage a beast that consumes them. Narrow Mouth Toads are equipped with a special adaptation to protect them from the ant, there being a flap of skin across the head that can voluntarily be moved forward to brush away any ants that attack the eyes.

Finally there are two species that seem to have disappeared from this Island in the past two or three years. They are the Florida Cricket Frog, *Acris gryllus dorsalis,* and the Florida Chorus Frog, *Pseudacris negrita verrucosa.* These little animals I believe to be extinct on Sanibel due to the overuse of organophosphate insecticides and to salt water intrusion. Perhaps they will appear again during a good rainy season, but I really doubt it.

Since the technical names of all amphibian species are included in the text, a check list will not follow this chapter.

The Giant Toad, ***Bufo marinus,*** *has invaded many lands.*

25 *CONSIDER SANIBEL'S OWN COTTON RAT* It Deserves A Better Deal

In our culture, rats are generally considered to be obnoxious, pestiferous creatures to the extent that the term "rat" is often used in a derogatory sense in the language.

For example, a person who is not receiving favor can be called a "rat." The word can be used as a verb in an uncomplimentary way, when someone "rats on" someone else. A foul thing may be said to "smell like a dead rat". Cats are considered "good" because they eat rats. Snarled, snagged long hair is "ratty." And a preacher or teacher might exclaim "Oh rats!" when a more familiar four-letter Americanism would be inappropriate.

Actually all this reflects a deep hate complex for rats which originates with a few species of truly obnoxious overconsuming and disease-carrying Asian rats that have spread all over the world. Collectively these few murid rodent species constitute the greatest and most costly animal pest problem in the world. The two most abundant of these Asian rats are called *Rattus rattus,* the Roof Rat, and *Rattus norvegicus* often called, quite mistakenly, the "Norway" Rat. "Mistakenly," because it originated ages ago in the steppes of northeast Asia, not Scandinavia.

The complete extermination of these two animals, if such *could* be accomplished, would probably solve the devastating food shortage problem that is confronting the human race today and which is undoubtedly going to result in additional widespread famine in Africa, Asia and Latin America in this century and perhaps in Europe and North America in the next.

The Roof Rat, also called the Black Rat, probably originated in what is today Indonesia as an arboreal forest dweller and was taken to Europe in the holds of cargo vessels. Even today it tends to live high above the ground in buildings or trees. Its transport system in our southern cities is the network of utility wires. The Norway Rat, often called the Brown Rat, tends to find shelter on or under the ground, probably due to its ancient origin in the treeless Steppe. It probably first reached Europe with invading central Asian Mongol hordes. It is this species that exists in huge populations in our northern cities in the sewer and other underground utility systems. These two species and their little cousin, the House Mouse, *Mus musculus,* all of which have invaded Sanibel, are far more successful than any other mammals in the history of the world and far outnumber the other "top" species in this country, man, by a factor of at least two. And they eat a lot. Up to 25% of all crops are either eaten or spoiled by murids. They present such an insurmountable problem that the FDA has even established acceptable levels of rat feces in certain human food products, thus officially recognizing the real dominance of these species.

So while the word rat has every good reason to be used in a derogatory way, it is unfortunate and unfair since our little native cricetid Cotton Rats don't look at all like the Asian or murid rats that are so troublesome throughout the world, but rather are quite attractive little beasts. Except for the tail, they look more like hamsters. They have quite beautiful pelage and are not obnoxious creatures at all. They are really woodland, scrub and marsh animals whose range covers much of North America, and in fact much of Central and South America as well. Some races are common in cotton fields of the Deep South, hence the common name.

The Sanibel animal is quite unique because it has evolved on these islands. Actually the first or "type" specimen was taken on adjacent Captiva which is connected to Sanibel by a short bridge at Blind Pass, and for this reason many students call it the Captiva Cotton Rat. Since

Sanibel Cotton Rat

it is rather more abundant on larger Sanibel, I've taken the liberty of calling it, at least insofar as its vernacular name is concerned, the "Sanibel Island Cotton Rat."

From the drawing, you can see a little animal that is not at all unpleasant and is truly an attractive addition to our fauna.

Luckily other people also feel this animal should be cared for and not destroyed. Dr. James N. Layne, Director of Research at the American Museum of Natural History's Archbold Biological Station, Lake Placid, Florida, has prepared a description of this animal which is presently listed in the "Provisional List of Rare and Endangered Plants and Animals." Consequently this endangered little animal is likely to receive legal protection. This might not be of much value unless those of us who live here in its habitat can recognize it and perhaps redirect some of the hate normally aimed at rats to an attitude of care for this rather choice and truly rare animal.

The Sanibel Cotton Rat, *Sigmodon hispidus insulicola,* has a rather chunky body, moderately long tail, relatively small ears and eyes, and long grizzled, bristly and quite attractive pelage. Its original habitat was in the upland ridge areas and adjacent swales, but now it occurs in various other habitats on Sanibel including the drier areas of the altered freshwater marsh, dry, open grassy fields and dry, grassy brush lands.

This animal was included on the endangered list because of the considerable development and habitat destruction that is taking place on Sanibel. It was thought that with its limited range there was some cause for concern for its future status. But it seems to be adapting to new habitats — more specifically to areas that have been developed — and it can now be found near residential areas. So it is likely to persist on the Island.

Of course, in the large areas of its natural habitat in the "Ding" Darling National Wildlife Refuge, and the not inconsiderable land areas of the Sanibel-Captiva Conservation Foundation, it will have an even better chance to survive on a permanent basis.

I had a very interesting experience with a female Sanibel Cotton Rat taken in a live trap. Too busy to handle her properly at the time, I simply provided some grain and gave her a water bottle and placed the trap in my lab to deal with later. That night, September 25th, she gave birth to a litter of five youngsters. Baby rodents of many kinds are born hairless and with their eyes closed and are quite helpless pink things. These little animals, however, proved to be precocious and within four days they had hair and their eyes were open. At five days they were eating solid foods and at six days they were separated from their mother, who was released. The youngsters were distributed to interested people who wished to raise them. I kept a pair for a month at which time they were almost full grown — very healthy, but very wild. They did not adapt well to cage life and were released.

There is another rodent here on Sanibel that is unique to our Island, having first been discovered here in 1955 and never found elsewhere. It is called the Sanibel Island Rice Rat, *Oryzomys palustris sanibeli.*

There are several other subspecies of Rice Rats ranging throughout the southeast from Texas and Oklahoma eastward and from Maryland southward. This is a marsh animal and is known to consume rice, hence its common name. On Sanibel it lives in the wetlands and feeds on grasses, roots and seeds.

Less attractive, to me at least, and more "ratty" than the Cotton Rat, the Rice Rat is grey-brown above, paler on the sides and whitish below. It has a longish tail and a pointed rat's nose. Nonetheless it is our very own subspecies and we had better appreciate it. The large Foundation wetland tracts should provide it with adequate permanent habitat.

Also a cricetid, the Rice Rat — unlike the Cotton Rat — does not seem to be doing so well. Perhaps it is also deserving of a place on the Endangered Species List. In three years of trapping effort, I have found this rare creature only a few times.

An ever present danger to both of these unusual native species is overcrowding of the habitat by the alien murids. Sample trapping that I have conducted in recent years has produced ninety percent murids to about ten percent natives.

AFTERWORD
Man—The Endangering Species

Who really cares whether tree frogs live on Sanibel? What difference does it really make if the Gopher Tortoise disappears? Otters, Crown Conchs, Alligators, Roseate Spoonbills, Olive Shells, Indigo Snakes — do they really make any difference in the quality of our anthropocentric lives?

You better believe they do, and here are a few reasons why:

Thousands of people take great pleasure in observing wildlife. Many who don't have the actual opportunity to see the rarer species still derive pleasure just from knowing that they exist.

Escapees from our urban malignancies renew and refresh their spirits by visits to beaches and sanctuaries such as those on Sanibel.

But there is a lot more to it than that. Disruptions of "the balance of nature" leave gaps in natural systems, sometimes causing all kinds of problems. For example, explosions of prey species, such as an overabundance of rodents, or hordes of insect pests, occur when their natural predators have been killed off. Or crashing predator populations can result from the elimination of their prey. When winged insects are killed off, some insectivorous birds may find themselves in trouble. The potential end result of such problems cannot even be imagined.

Much of the natural world remains a mystery today. Who knows what startlingly important discoveries lie in the future? Suppose, for example, bread mold had been eliminated before Fleming discovered penicillin. Or consider the problems that would have resulted if an obscure grass species in Guatemala had been destroyed before the maize hybridizers were able to produce today's high-yield corn. All living things must be preserved for future study. The loss of a single organism through human agency is intolerable.

Regretably, one of man's most enduring activities has been the rendering of Planet Earth poorer in life forms. What can be more permanent than the extinction of a species or the elimination of natural systems like those found on this sanctuary Island? How can man in good conscience attach *sapiens* to his name and at the same time continue his pattern of destruction?

Let all of us who know and love Sanibel work together to preserve her natural values. Let this Island be a model in microcosm of man's ability to preserve and restore a place of great natural beauty and value. Let us show the world attitudes and principles worthy of emulation. In some areas of human activity, "Sanibel" is synonymous with "leadership." Let this be true also in the practice of the fine art of environmental stewardship. Let us learn to live with nature and no longer attempt to dominate her. Then, perhaps, we will have earned the right to call ourselves *Homo sapiens.*

APPENDIX I

A TENTATIVE CHECKLIST OF THE MAMMALS OF SANIBEL ISLAND, LEE COUNTY, FLORIDA

This list contains all mammal species known to occur on Sanibel; others known from the recent past, now extinct; and still others that may occur since they can be found on adjacent mainland areas or other nearby barrier islands. Marine mammals known to exist or to have existed in the waters adjacent to Sanibel are also listed. An asterisk (*) indicates an introduced taxon. An upward arrow (↗) indicates the form is increasing, a horizontal arrow (→) means holding steady, while a downward-pointing arrow (↘) denotes decline. For some, no arrow is shown. For these no trend judgments have been attempted.

Of the thirty-two taxa, or kinds, for which trend judgments have been attempted, I feel that seven are increasing, twenty-one are declining or have disappeared, and four are holding their own.

A significant number of forms are, deplorably, declining due to:

1. Habitat destruction — clearing, salt intrusion, draining, filling and other development.
2. Ever-increasing high-speed automobile traffic.
3. Continued abundant use of organophosphate and other pesticides.
4. Other pressures meted out by the overabundant dominant primate species.

Order MARSUPIALIA — the Marsupials

Didelphidae — Opossum family

↗ *Didelphis virginiana pigra** — Opossum (Introduced to Sanibel from the mainland after the bridge was built.)

Order INSECTIVORA — the Insectivores

Soricidae — Shrew family

↘ *Blarina brevicauda* — Short-tailed Shrew — known only from North Fort Myers

↘ *Cryptotis parva floridana* — Florida Least Shrew — throughout peninsular Florida

N. B. Neither of these shrews has been collected on Sanibel, but the search continues as both can very possibly be present.

Talpidae — Mole family

↘ *Scalopus aquaticus* — Eastern Mole

Order CHIROPTERA — the Bats

Vespertilionidae — Vespertilionid Family

→ *Lasiurus intermedius* — Florida Yellow Bat — known on Sanibel (1974)

↘ *Myotis austroriparius* — Southeastern Myotis

↘ *Nycticeius humeralis* — Evening Bat

(Not verified but probably on Sanibel)

Molossidae — Molossid Bat Family

↘ *Tadarida brasiliensis cynocephala* — Brazilian Free-tailed Bat. Probably on Sanibel, as known from Marco.

Order PRIMATA — the Primates

Cebidae — Prehensile-tailed Monkey Family

Cebus sp. — Capuchin Monkey* (escapees from primate collections)

Hominidae — Ape and Man Family

Homo sapiens americanus (Linn.) — Caloosa — Probably extinct

Homo sapiens afer (Linn.) — African Man*

Homo sapiens asiaticus (Linn.) — Asian Man*

Homo sapiens sapiens (Linn.) — Modern Man* — abundant and destructive on Sanibel.

Order EDENTATA — the Edentates

Dasypodidae — the Armadillo Family

Dasypus novemcinctus mexicanus — Nine-banded Armadillo* — well-known on Sanibel.

Order LAGOMORPHA — the Lagomorphs

Leporidae — Hare and Rabbit family

Sylvilagus palustris paludicola — Marsh Rabbit — abundant on Sanibel.

Order RODENTIA — the Rodents

Sciuridae — Squirrel Family

Sciurus carolinensis extimus — Gray Squirrel

Cricetidae — New World Rat and Mouse Family

Oryzomys palustris sanibeli — Sanibel Island Marsh Rice Rat. Few observed since 1974; rare.

Peromyscus gossypinus — Cotton Mouse — Known on barrier islands, Southwest Florida; may be on Sanibel, but not confirmed.

Peromyscus floridanus — Florida Mouse — on scrub palmetto ridges. May be on Sanibel, but not confirmed.

Sigmodon hispidus insulicola — Sanibel Hispid Cotton Rat — an endangered animal, unique to Sanibel and Captiva. Also thought to be on Pine Island, Little Pine Island and possibly at Englewood.

Microtinae — subfamily — The Microtine Rodents

Neofiber alleni nigrescens — Round-tailed Muskrat — known on Sanibel.

Muridae — Old World Rat and Mouse Family

Mus musculus brevirostris — House Mouse* — common on Sanibel

Rattus rattus — Roof Rat* — Known to be abundant at Lighthouse

Rattus norvegicus — Norway Rat* — Not abundant on Sanibel, but observed.

Order CETACEA — the Cetaceans

Ziphiidae — Beaked Whale Family

Mesoplodon gervaisi — Gervais' Beaked Whale

Ziphius cavirostris — Goose-beaked Whale

Delphinidae — Porpoise and Dolphin Family

Delphinus delphis — Atlantic Dolphin

Orcinus orca — Atlantic Killer Whale

Pseudorca crassidens — False Killer Whale

Stenella frontalis — Cuvier's Porpoise

Stenella plagiodon — Spotted Porpoise

Tursiops truncatus — Atlantic Bottle-nosed Dolphin

Globicephala macrorhyncha — Pilot Whale

Balaenopteridae — Fin-backed Whale Family

Balaenoptera acutorostrata — Little Piked Whale

Balaenidae — Right and Bowhead Whale Family

Eubalaena glacialis — Atlantic Right Whale

The above whale species have been observed at one time or another in waters adjacent to Sanibel. There are about an equal number of additional cetacean taxa that are probable for this region.

Order CARNIVORA — the Carnivores

Canidae — the Dog Family

Urocyon cinereoargenteus floridanus — Gray Fox — known from Sanibel 1974 and 1975

Canis familiaris — Domestic Dog*

Ursidae — the Bear Family

Euarctos floridanus — Florida Black Bear — probable past Sanibel resident or visitor

Procyonidae — the Raccoon Family

Procyon lotor — Raccoon — abundant on Sanibel

Mustelidae — the Weasel Family

Lutra canadensis — River Otter — well known on Sanibel

Mephitis mephitis — Striped Skunk — known on Sanibel, but rare

Mustela frenata peninsulae — Florida Long-tailed Weasel — not known, but probably on Sanibel

Mustela vison, subspecies — Mink — known in Darling refuge and on Captiva

Spilogale putorius ambarvalis — Eastern Spotted Skunk — known on Sanibel but rare

Felidae — Cat Family

Felis concolor coryi — Florida Panther

Felis (sometimes *Lynx*) *rufus floridanus* — Florida Bobcat — Present resident on Sanibel

Felis catus — Feral Domestic Cat* — many currently living on Sanibel and causing serious problems.

Order PINNIPEDIA — the Pinnipeds

Otariidae — Sea Lion Family

Zalophus californianus — California Sea Lion* — Sighted many times in this region.

Phocidae — Earless Seal Family

Monachus tropicalis — West Indian Seal — Now extinct in most or all of its range, however certainly a past Sanibel visitor.[1]

Order SIRENIA — the Manatees and Dugongs

Trichechidae — the Manatee Family

Trichechus manatus latirostris — West Indian Manatee — Some observed every year. Many cut by boat propellers. Also victim of pollution.

Order ARTIODACTYLA — Even-toed Ungulates

Cervidae — the Deer Family

➘ *Odocoileus virginiana* — White-tailed Deer — Became extinct after World War I on Sanibel but well-known here early in this century. There is now evidence of their return.

Suidae — the Pig Family

➘ *Sus scrofa* — Feral Domestic Swine* — Known in the past on Sanibel, but now believed to be completely eliminated.

---

[1]Recent excavations of a Caloosa Indian Shell Mound, the lower levels of which were a pre-Caloosa campsite, definitely reveal the presence of *Monachus* at Sanibel, according to A. Fradkin. Dr. C. J. Wilson, Sanibel anthropologist was in charge of the "dig" and he, too, confirmed to me that *Monachus* was a probable past resident.

APPENDIX II

The Code of Ethics
of the
AMERICAN MALACOLOGICAL UNION

"The many thousands of living species of mollusks are significant components of Earth's web of life. They are found in oceans, lakes, rivers, prairies, mountains and even deserts. Mollusks aid in the natural recycling of plant and animal matter to new forms of life. They serve as food for many larger animals, including humans. Many mollusks produce shells and other products which have been admired and used by man for millenia.

Each species has its own unique attributes which enable its members to survive and reproduce in a specialized environmental niche. The great diversity of species of mollusks contributes resilience and stability to the world's ecosystem.

The American Malacological Union, a society of professional and amateur malacologists, is much concerned by the continuing gradual depletion of the world's natural resources of molluscan species. It, therefore, supports measures being taken to prevent further environmental degradation and recommends the establishment of ecological preserves for the perpetuation of natural biological communities, especially those containing species in danger of extinction.

The American Malacological Union urges all who are concerned throughout the world to accept responsibility for ensuring the future existence of mollusks and their habitats.

We, the members of the American Malacological Union, therefore accept and endorse the following guidelines for our field studies and collecting activities:

1. Observation and photography of mollusks in their natural habitats can yield important biological information and is often a more rewarding activity than the collecting of living animals. The A.M.U. encourages such observational research by both amateur and professional malacologists.
2. Living specimens should be collected only in those minimal numbers necessary to satisfy the requirements of the study. Dead shells often make valuable specimens, and their collection does not further endanger the population. The A.M.U. encourages the collection of dead shells, especially in cases where the soft parts are not required for anatomical or physiological research.
3. Because detailed, properly documented material is needed to establish the ranges and habitats of all molluscan species and to ensure the success of efforts to conserve these animals, the A.M.U. urges all collectors to carefully label all specimens, photographs and field notes with the precise locality, the exact date and the full name of the observer. It further recommends that arrangements be made for the deposition of such documentation and specimens in permanent museum reference collections for study by other malacologists when the original studies are completed.
4. The results of field studies should be shared as widely as possible by means of educational exhibits, published paper, letter, seminars and lectures.
5. The laws concerning collecting and trespass are to be known and obeyed by all. Field workers will obtain all necessary licenses and permits from official agencies and landowners before engaging in collecting or other activities.

In order to advance man's knowledge of the molluscan fauna of the Earth, and in the spirit of conservation of these natural resources, the American Malacological Union hereby signifies its intent to adhere to these guidelines."

The above Code of Ethics was written by Dr. Carol B. Stein and adopted at the Annual Business Meeting of the A.M.U. held on June 27, 1973. It is quoted by permission.

THE NATIONAL ARCHIVES
LITTERA SCRIPTA MANET

FEDERAL REGISTER

OF THE UNITED STATES
1934

VOLUME 12 NUMBER 235

Washington, Wednesday, December 3, 1947

TITLE 3—THE PRESIDENT

PROCLAMATION 2758

CLOSED AREA UNDER THE MIGRATORY BIRD TREATY ACT—FLORIDA

BY THE PRESIDENT OF THE UNITED STATES OF AMERICA

A PROCLAMATION

WHEREAS the Acting Secretary of the Interior has submitted to me for approval the following regulation adopted by him, after notice and public procedure pursuant to section 4 of the Administrative Procedure Act of June 11, 1946 (60 Stat. 238), under authority of the Migratory Bird Treaty Act of July 3, 1918 (40 Stat. 755, 16 U. S. C. 704), and Reorganization Plan No. II (53 Stat. 1431):

REGULATION DESIGNATING AS CLOSED AREA CERTAIN LANDS AND WATERS WITHIN, ADJACENT TO, OR IN THE VICINITY OF THE SANIBEL NATIONAL WILDLIFE REFUGE, FLORIDA

By virtue of and pursuant to the authority contained in section 3 of the Migratory Bird Treaty Act of July 3, 1918 (40 Stat. 755, 16 U. S. C. 704), Reorganization Plan No. II (53 Stat. 1431), and in accordance with the provisions of section 4 of the Administrative Procedure Act of June 11, 1946 (60 Stat. 238), I, Oscar L. Chapman, Acting Secretary of the Interior, having due regard to the zones of temperature and to the distribution, abundance, economic value, breeding habits, and times and lines of flight of the migratory birds included in the terms of the conventions between the United States and Great Britain for the protection of migratory birds, concluded August 16, 1916, and between the United States and the United Mexican States for the protection of migratory birds and game mammals, concluded February 7, 1936, do hereby designate as closed area, effective thirty days after publication in the FEDERAL REGISTER, in or on which pursuing, hunting, taking, capturing, or killing of migratory birds, or attempting to take, capture, or kill migratory birds is not permitted, all areas of land and water in Lee County, Florida, not now owned or controlled by the United States within the following-described exterior boundary:

Beginning at low water east of Sanibel Island Light, situated on Point Ybel on the east end of Sanibel Island, Florida, in approximate latitude 26°27'13" N., longitude 82°00' 48" W.;

Thence Northwesterly with low water along the northeast side of Sanibel Island approximately 5170 yards (2.94 miles) to a point at low water and approximately 704 yards (0.40 mile) southeast of Woodrings Point on Sanibel Island;

Thence Northwesterly, within Pine Island Sound, approximately 1760 yards (1.00 mile) to St. James Light 5, in St. James City Channel, between Sanibel and Pine Islands;

Thence Northwesterly, Southwesterly and then Northwesterly continuing through Pine Island Sound by straight lines connecting in order the following navigation markers: St. James Daybeacon 7 (black); Pine Island Sound Daybeacon 8 (red); Pine Island Sound Light 10; Pine Island Sound Daybeacon 12 (red); Pine Island Sound Daybeacon 14 (red); Pine Island Sound Daybeacon 15 (black); Pine Island Sound Light 16, approximately 11,616 yards (6.60 miles) to Wulfert Daybeacon 1 (black), at the entrance to Wulfert Channel;

Thence Westerly and Southwesterly in Wulfert Channel and between Sanibel and Captiva Islands by straight lines connecting in order the following navigation markers: Wulfert Daybeacon 3 (black); Wulfert Daybeacon 5 (black); Wulfert Daybeacon 7 (black); Horn Passage Daybeacon 2 (red), approximately 3058 yards (1.74 miles) to Horn Passage Daybeacon 3 (black);

Thence Southerly approximately 638 yards (0.36 mile) to the center of the highway bridge connecting Sanibel and Captiva Islands;

Thence Westerly with the center of bridge and the prolongation thereof, across Captiva Island approximately 506 yards (0.29 mile) to low water on the west shore of Captiva Island;

Thence Westerly at right angles to the shore of Captiva Island, 440 yards (0.25 mile) to a point in the Gulf of Mexico;

Thence Southeasterly, Easterly and then Northeasterly, in the Gulf of Mexico, parallel to and 440 yards (0.25 mile) from low water along the south shore of Captiva and Sanibel Islands approximately 24,024 yards (13.65 miles) to a point in the Gulf of Mexico;

Thence Northwesterly 440 yards (0.25 mile) to the place of Beginning.

IN WITNESS WHEREOF, I have hereunto subscribed my name and caused the seal of the Department of the Interior to be affixed this tenth day of November 1947.

[SEAL] OSCAR L. CHAPMAN,
Acting Secretary of the Interior.

(Continued on p. 8041)

AND WHEREAS upon consideration it appears that the foregoing regulation will tend to effectuate the purposes of the aforesaid Migratory Bird Treaty Act of July 3, 1918:

NOW, THEREFORE, I, HARRY S. TRUMAN, President of the United States of America, under and by virtue of the authority vested in me by the aforesaid Migratory Bird Treaty Act of July 3, 1918, do hereby approve and proclaim the foregoing regulation of the Acting Secretary of the Interior.

IN WITNESS WHEREOF I have hereunto set my hand and caused the Seal of the United States of America to be affixed.

DONE at the City of Washington this 2nd day of December in the year of our Lord nineteen hundred and [SEAL] forty-seven, and of the Independence of the United States of America the one hundred and seventy-second.

HARRY S. TRUMAN

By the President:

ROBERT A. LOVETT,
Acting Secretary of State.

[F. R. Doc. 47-10701; Filed, Dec. 2, 1947; 12:08 p. m.]

This is President Truman's "Sanibel Closure Order". It includes the whole of Sanibel Island and in a sense it makes the whole Island a nature sanctuary. In 1947, the newsprint used in the Federal Register was of such poor quality that existing copies from that date fragment easily or are hard to read. Hence it is of value to copy this document here even though it is badly blurred.

INDEX

Molly Eckler B. whose fine drawings appear throughout this book is an outstanding young artist who has lived and worked on Sanibel and Captiva for the past five years. Her work has been widely acclaimed and she is one of the featured artists at Sanibel's Schoolhouse Gallery.

Molly has had great impact on the preservation of natural habitats in Southwest Florida. It was she who kicked off the effort that resulted in the State of Florida's acquisition of Cayo Costa, a barrier island that lies to the north of Sanibel and which will now be preserved in its natural state forever.

Molly graduated from Ohio Wesleyan University in 1969.

Ms. Eckler wishes to acknowledge the invaluable resource of Michael LaTona's photographs for the execution of her bird drawings, and the cooperation of C.R.O.W. in permitting observation of their animal patients. Her thanks, also, for the valuable assistance of Island Printing Centre, Photo Sanibel and Mark Williams of **Island Reporter**.

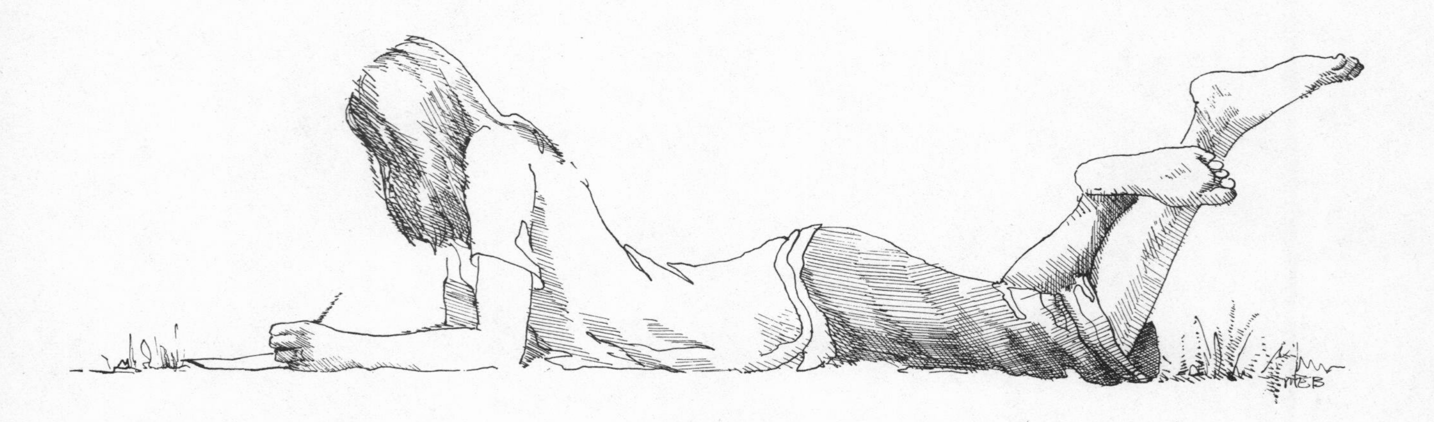